C4004

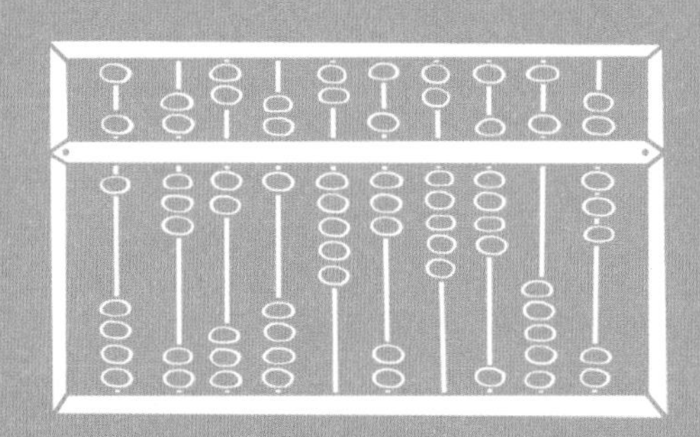

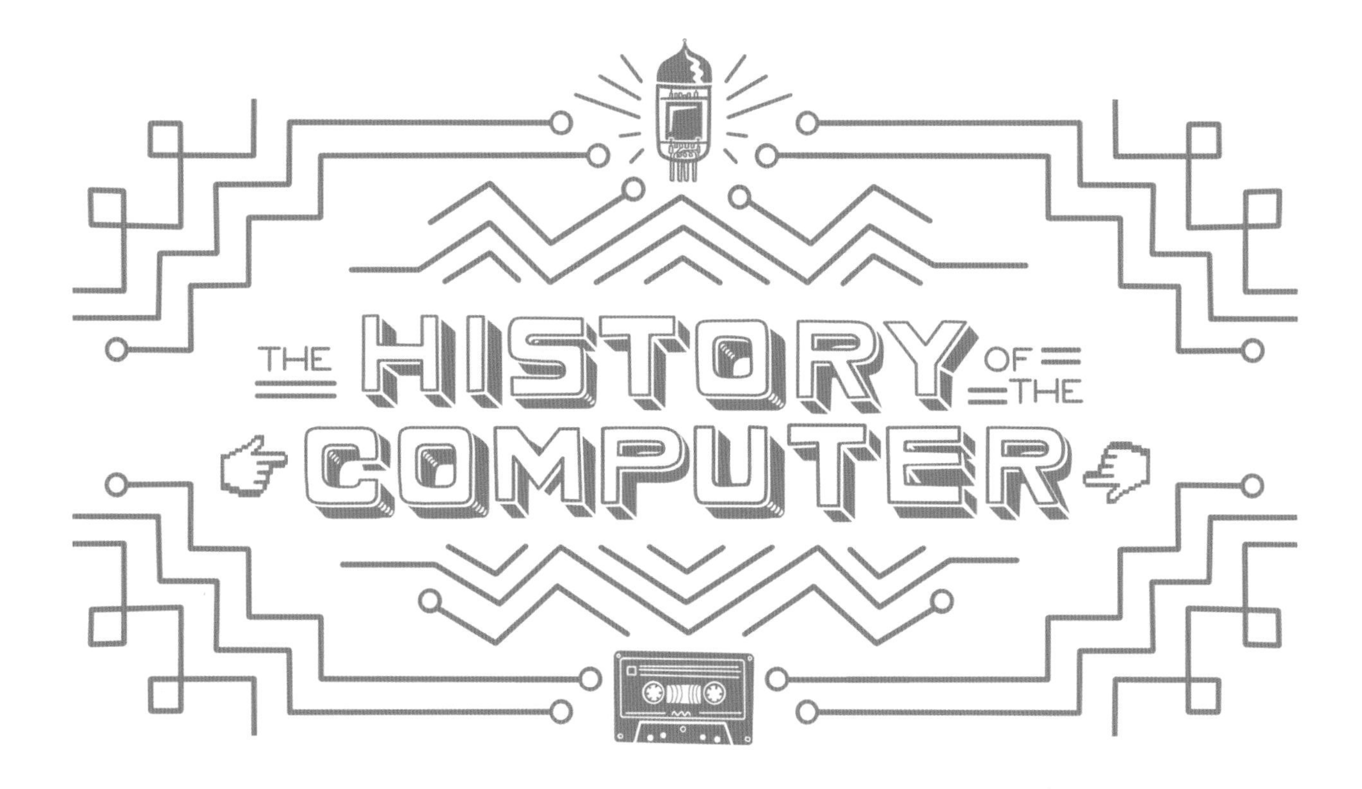
THE HISTORY OF THE COMPUTER

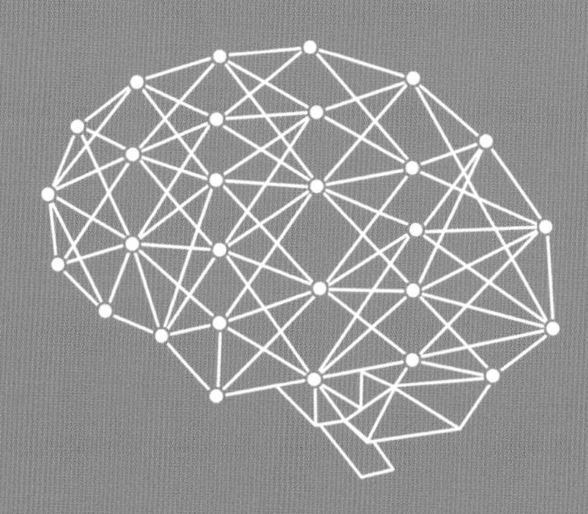

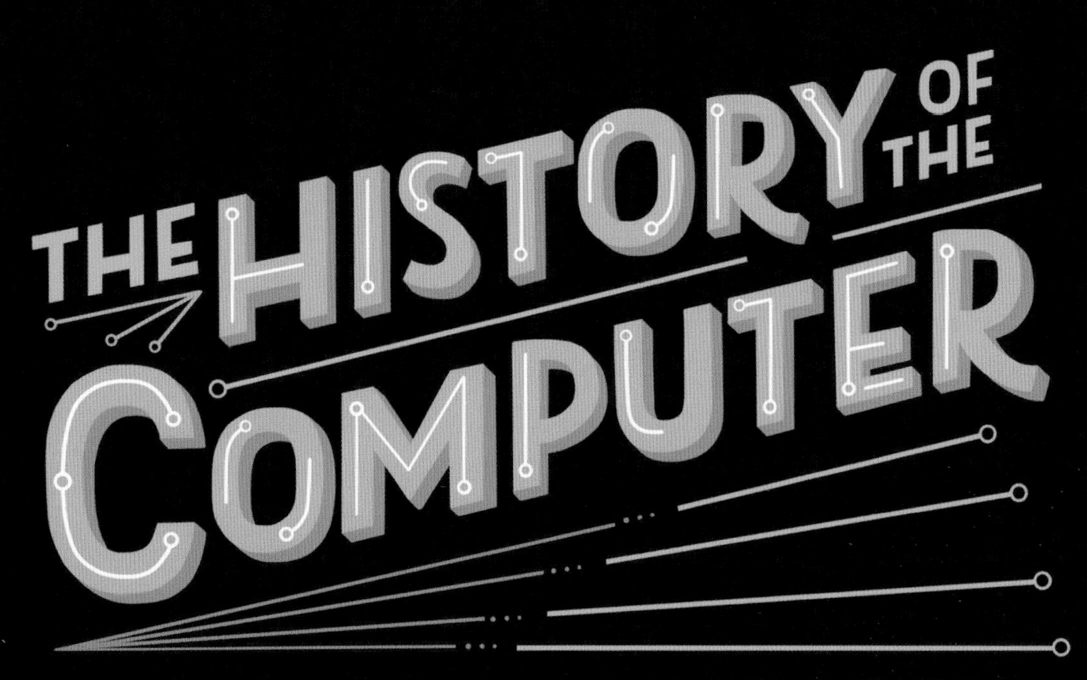

PEOPLE, INVENTIONS, AND TECHNOLOGY THAT CHANGED OUR WORLD

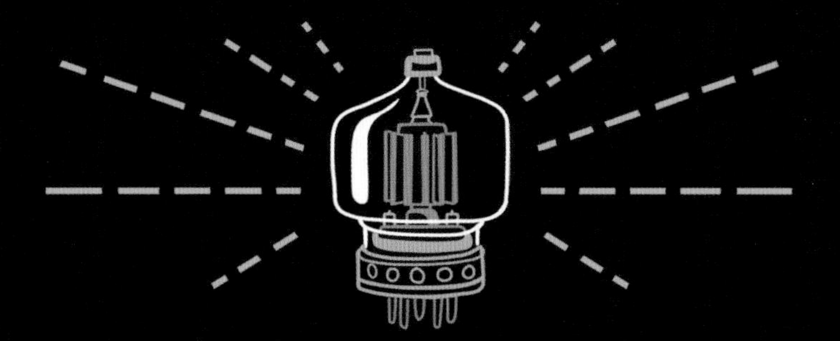

WRITTEN AND ILLUSTRATED BY

RACHEL IGNOTOFSKY

TEN SPEED PRESS
CALIFORNIA | NEW YORK

CONTENTS

INTRODUCTION · · · 6
INSIDE A COMPUTER · · · 8
BINARY AND ON/OFF SWITCHES · · · 10
MEMORY AND STORAGE · · · 12
VIDEO GAMES · · · 14
AI AND ROBOTS · · · 16

ANCIENT CIVILIZATIONS (25,000 B.C.E.–1599 C.E.) · · · 19
TIMELINE · · · 20
STORIES FROM HISTORY · · · 22
IMPORTANT INVENTIONS · · · 24
THE ABACUS AROUND THE WORLD · · · 26

STEAM AND MACHINES (1600–1929) · · · 29
TIMELINE · · · 30
STORIES FROM HISTORY · · · 32
IMPORTANT INVENTIONS · · · 36
INFLUENTIAL PEOPLE · · · 38

WWII AND THE FIRST COMPUTERS (1930–1949) · · · 41
TIMELINE · · · 42
STORIES FROM HISTORY · · · 44
IMPORTANT INVENTIONS · · · 48
INFLUENTIAL PEOPLE · · · 50

THE POSTWAR BOOM AND THE SPACE RACE (1950–1969) · · · 53
TIMELINE · · · 54
STORIES FROM HISTORY · · · 56
IMPORTANT INVENTIONS · · · 60
INFLUENTIAL PEOPLE · · · 62

THE PERSONAL COMPUTER (1970–1979) · · · 65
TIMELINE · · · 66
STORIES FROM HISTORY · · · 68
IMPORTANT INVENTIONS · · · 72
INFLUENTIAL PEOPLE · · · 74

COMPUTERS AS A CREATIVE TOOL (1980–1989) · · · 77
TIMELINE · · · 78
STORIES FROM HISTORY · · · 80
IMPORTANT INVENTIONS · · · 84
INFLUENTIAL PEOPLE · · · 86

THE WORLD WIDE WEB (1990–2005) · · · 89
TIMELINE · · · 90
STORIES FROM HISTORY · · · 92
IMPORTANT INVENTIONS · · · 96
INFLUENTIAL PEOPLE · · · 98

THE ALL-IN-ONE DEVICE (2006–NOW) · · · 101
TIMELINE · · · 102
STORIES FROM HISTORY · · · 104
IMPORTANT INVENTIONS · · · 110
INFLUENTIAL PEOPLE · · · 112

CHALLENGES IN A DIGITAL WORLD · · · 114
THE FUTURE · · · 116
CONCLUSION · · · 118

SOURCES AND RESOURCES · · · 120
ACKNOWLEDGMENTS · · · 122
ABOUT THE AUTHOR · · · 122
INDEX · · · 124

INTRODUCTION

"An Electronic Brain!" That's what a reporter on live national television called the UNIVAC—right before it performed the task of predicting the results of the 1952 presidential election. It was a risky publicity stunt, and no one, not even the engineers who had constructed this computer, knew what it would do. Beamed into homes nationwide on black-and-white TVs, this was the first time that the public had seen a computer in action. The computers that came before were gigantic, noisy machines, built during World War II and hidden away in top-secret laboratories. The UNIVAC (short for UNIVersal Automatic Computer) was new—built for offices instead of battle—and would now have to prove itself. The American public watched the UNIVAC's giant blinking console and rows of spinning magnetic tape while it did its calculations. Contrary to the polls, UNIVAC called a landslide victory for Dwight D. Eisenhower. To everyone's surprise, the UNIVAC was correct! It was sensational! The American audience was enthralled—it was sci-fi in real life! This TV event propelled computers into pop culture and the public imagination.

Computers have come a long way since 1952! We now have access to the entire sum of human knowledge with devices that fit in the palms of our hands. The inventions that have gotten us to this point are part of a technological journey that starts as far back as the Stone Age. Before we dive into the history of computers, let's define what a computer is.

A COMPUTER IS A MACHINE THAT STORES, RETRIEVES, AND PROCESSES DATA BY FOLLOWING A SET OF INSTRUCTIONS.

At its core, a computer is a tool that expands the capacity of the human mind. We are all familiar with the idea that tools enhance our physical abilities to do more work—a hammer helps our arm strike a nail. A computer is a tool that enhances our mental abilities. Computers help us solve complicated math equations, store and sort vast libraries of information, and even help us find our new favorite restaurant!

The internet, combined with the creation of the World Wide Web in 1990, turned computers into media machines. The web is an integral part of the global economy and, for many, an extension of a personal identity. Computers have become so integrated into our lives that, in 2011, the United Nations declared access to the internet a human right.

Billions of people have smartphones that are 100,000 times more powerful than the computer that flew the first astronauts to the moon. But a computer in nearly every pocket has not always been the case. For the majority of history, computing machines were used by very few people—scientists doing research, governments managing bureaucracies, militaries fighting wars, and large corporations looking to maximize their profits. Early computers were incredibly expensive and physically massive and required specialized technical knowledge to use. It wasn't until the personal computer revolution of the 1970s that the power of computing became more accessible to the average person.

This book will highlight the milestones of computer history and explore the idea that technological knowledge is power. It will not teach you how to code or go into the weeds of computer science. Instead, this book focuses on the intent, purpose, and impact of the people and machines that changed our world. While this book profiles innovators in technology, there is no such thing as a "lone genius." The origin of computers came from the work of thousands of people and a social climate that prioritized the funding of scientific advancement. The study of computer history is the study of humanity.

INSIDE A COMPUTER

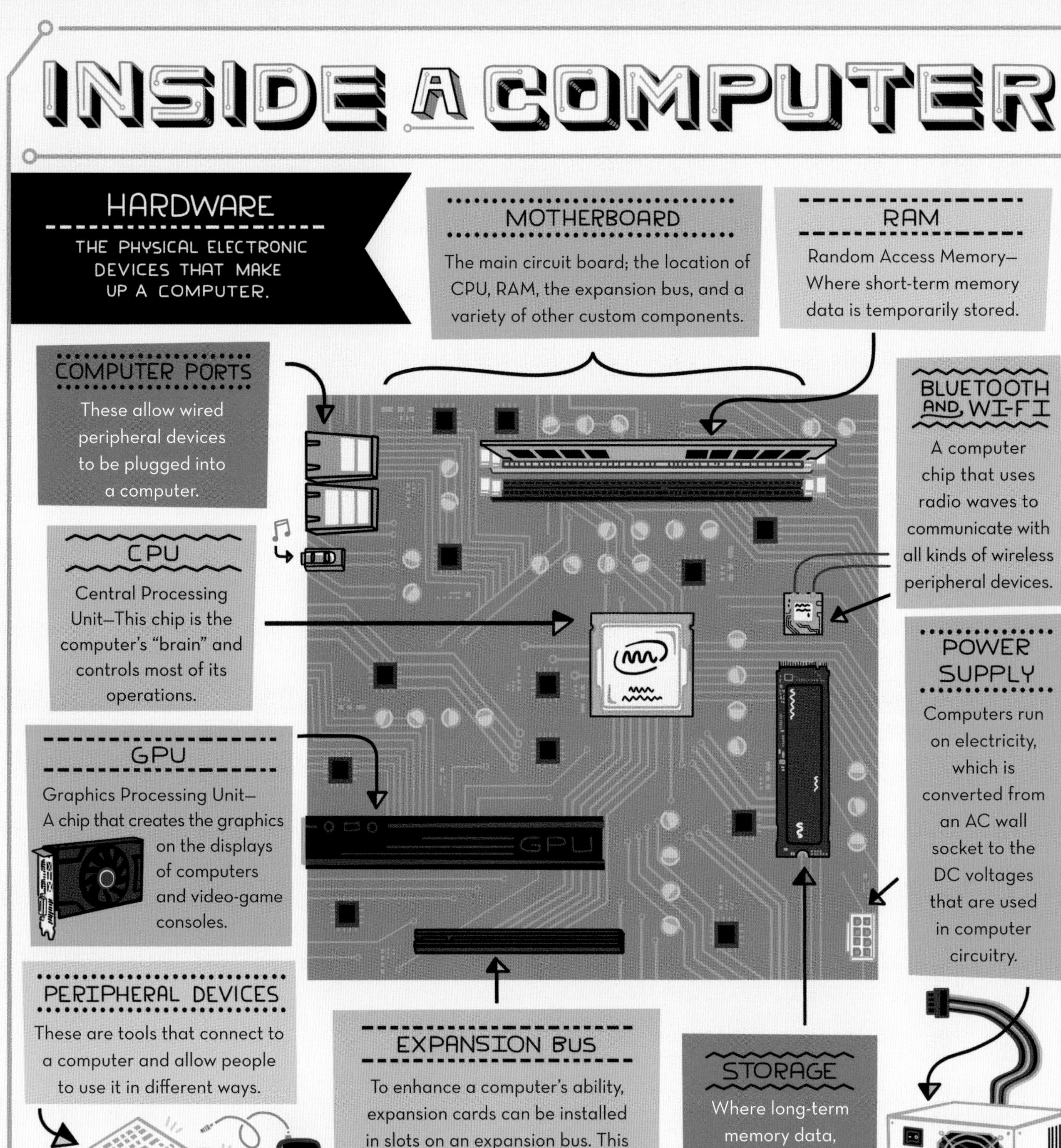

HARDWARE

THE PHYSICAL ELECTRONIC DEVICES THAT MAKE UP A COMPUTER.

MOTHERBOARD

The main circuit board; the location of CPU, RAM, the expansion bus, and a variety of other custom components.

RAM

Random Access Memory—Where short-term memory data is temporarily stored.

COMPUTER PORTS

These allow wired peripheral devices to be plugged into a computer.

BLUETOOTH AND WI-FI

A computer chip that uses radio waves to communicate with all kinds of wireless peripheral devices.

CPU

Central Processing Unit—This chip is the computer's "brain" and controls most of its operations.

POWER SUPPLY

Computers run on electricity, which is converted from an AC wall socket to the DC voltages that are used in computer circuitry.

GPU

Graphics Processing Unit—A chip that creates the graphics on the displays of computers and video-game consoles.

PERIPHERAL DEVICES

These are tools that connect to a computer and allow people to use it in different ways.

EXPANSION BUS

To enhance a computer's ability, expansion cards can be installed in slots on an expansion bus. This allows for extra storage, more GPUs, and new ports for connecting to all kinds of hardware.

STORAGE

Where long-term memory data, including software and files saved by the user, is stored.

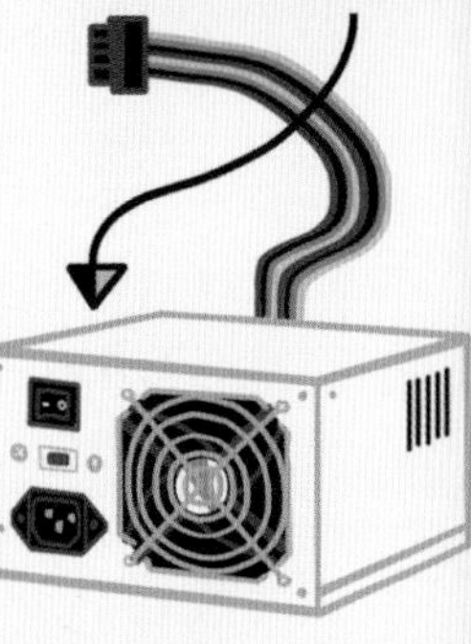

SOFTWARE TERMS

SOFTWARE

The programs that tell a computer what to do, such as applications, operating systems, and firmware.

COMMAND LINE

This is a technical way to operate a computer using only text-based commands. It uses very little resources, like memory or processing power, and was the dominant way that people interacted with computers until the 1990s.

GUI

Graphic User Interface—A graphic representation of files and software that allows the user to easily instruct the computer what to do. Modern operating systems use icons and graphics that are easy to understand. When interacting with a computer, the GUI puts on a "theater show" to communicate with the user what the computer is doing.

OS

Operating System—This software is like the conductor of an orchestra. It manages a computer's hardware and software and makes it easy to use. Over the years, there have been many different operating systems, and people generally have personal favorites. Examples include Linux, MacOS, Windows, and Android.

PROGRAM

Coded instructions for a computer to execute a specific task.

PROGRAMMING LANGUAGES

Computers understand only binary (a.k.a. machine) code. Binary is hard for people to understand, so programmers use programming languages to create software. These complex languages are then translated back into binary code (by a compiler or assembler), which the computer understands.

COMPUTERS MERGE TECHNOLOGIES

Throughout history, computers have merged all kinds of different technologies to create something new. A great example of this is how, over time, cameras have become an essential part of modern devices.

Smartphones combined the technology of desktop computers, telephones, touch screens, and GPS (Global Positioning System) to create an "all-in-one device." When someone reaches for a "camera" to take a photo, they often grab this pocket-size computer.

BINARY AND ON/OFF SWITCHES

1s AND 0s

Classical computers understand only the electrical signals of ON and OFF. To "talk" to a computer, people use binary code, which is also called machine code. *Binary* means "of two states," and binary code is made of 1s and 0s. ON is represented by 1, and OFF is represented by 0. Every operation performed by a computer and every single piece of data that is processed by a computer are represented by 1s and 0s.

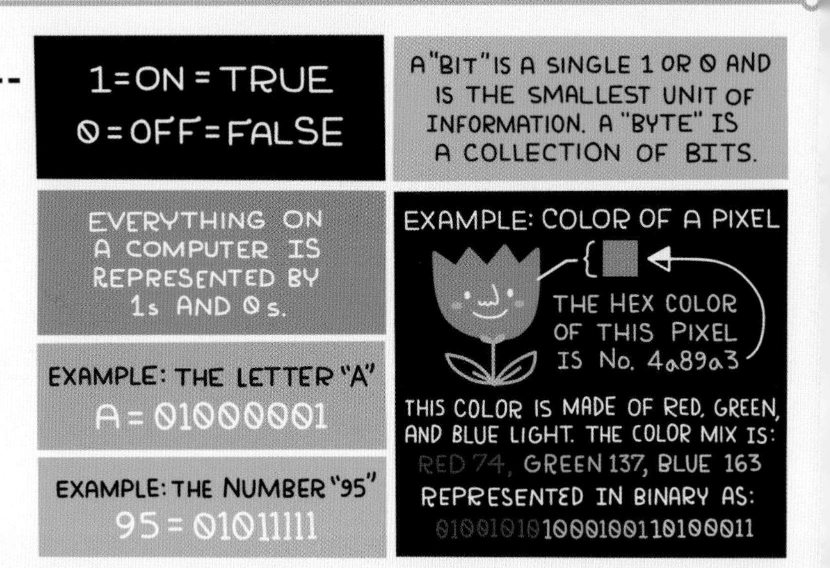

BOOLEAN ALGEBRA AND LOGIC GATES

In 1847, mathematician George Boole developed the rules for Boolean algebra, which is a branch of mathematics that is all about determining whether a logical statement is true or false, using the operations NOT, AND, and OR. The binary of true and false can also be represented by a 1 or a 0 and easily translates to an electrical current being ON or OFF. In 1937, electrical engineer Claude Shannon realized he could represent Boolean logical statements physically with electrical circuits, using ON/OFF switches. These are called logic gates. Shannon is considered one of the founders of digital circuit design.

Electricity passes through the logic gates in a computer's circuitry like water through a system of pipes. Think of the ON/OFF switches as a faucet, directing the flow of electricity. Logic gates manipulate electrical signals that represent 1s and 0s. Simple logic gates combine to create the complex architecture inside a computer chip.

THE 3 FUNDAMENTAL OPERATIONS IN BOOLEAN ALGEBRA

TRUTH TABLE 1=TRUE 0=FALSE

NOT — THIS IS A NEGATION.

IF THE INPUT IS TRUE, THEN THE OUTPUT IS NOT TRUE (IT'S FALSE) AND VICE VERSA.

INPUT	OUTPUT
1	0
0	1

AND — THIS IS A CONJUNCTION

THE OUTPUT IS ONLY TRUE WHEN BOTH INPUT STATEMENTS A AND B ARE TRUE.

INPUT A	INPUT B	OUTPUT
0	0	0
0	1	0
1	0	0
1	1	1

OR — THIS IS A DISJUNCTION

THE OUTPUT IS TRUE IF EITHER INPUT STATEMENTS A OR B (OR BOTH) ARE TRUE.

INPUT A	INPUT B	OUTPUT
0	0	0
0	1	1
1	0	1
1	1	1

OTHER LOGIC OPERATIONS INCLUDE XOR (EXCLUSIVE OR), XNOR (EXCLUSIVE NOT OR), NAND (NOT AND) & NOR (NOT OR).

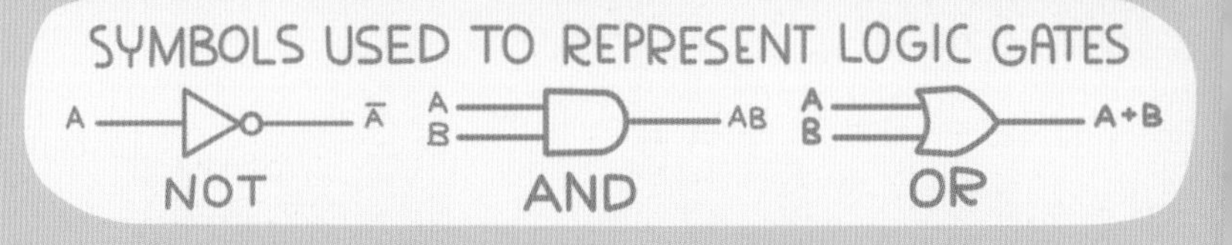

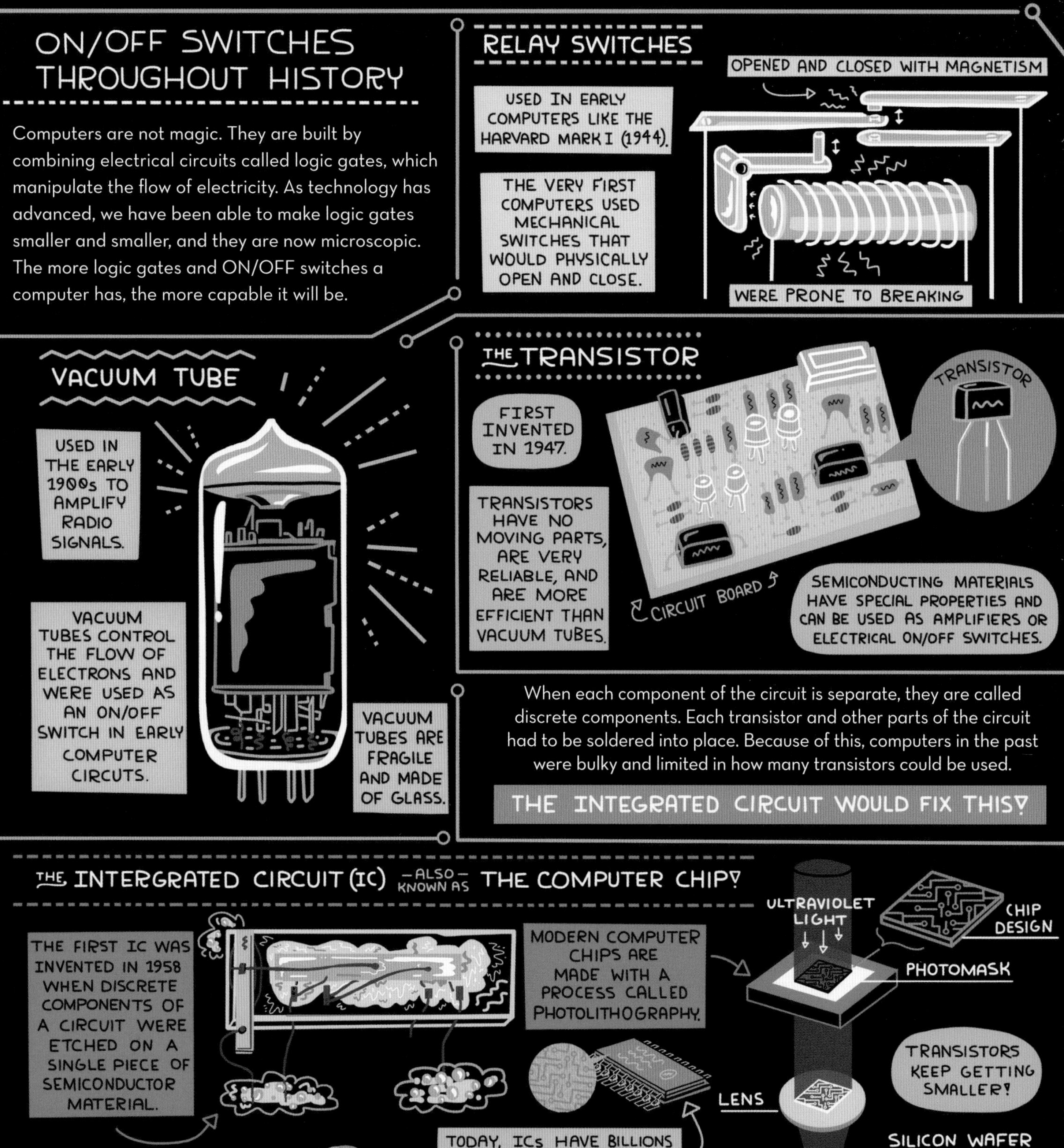
ON/OFF SWITCHES THROUGHOUT HISTORY
Computers are not magic. They are built by combining electrical circuits called logic gates, which manipulate the flow of electricity. As technology has advanced, we have been able to make logic gates smaller and smaller, and they are now microscopic. The more logic gates and ON/OFF switches a computer has, the more capable it will be.
RELAY SWITCHES
USED IN EARLY COMPUTERS LIKE THE HARVARD MARK I (1944).
THE VERY FIRST COMPUTERS USED MECHANICAL SWITCHES THAT WOULD PHYSICALLY OPEN AND CLOSE.
OPENED AND CLOSED WITH MAGNETISM
WERE PRONE TO BREAKING
VACUUM TUBE
USED IN THE EARLY 1900s TO AMPLIFY RADIO SIGNALS.
VACUUM TUBES CONTROL THE FLOW OF ELECTRONS AND WERE USED AS AN ON/OFF SWITCH IN EARLY COMPUTER CIRCUTS.
VACUUM TUBES ARE FRAGILE AND MADE OF GLASS.
THE TRANSISTOR
FIRST INVENTED IN 1947.
TRANSISTORS HAVE NO MOVING PARTS, ARE VERY RELIABLE, AND ARE MORE EFFICIENT THAN VACUUM TUBES.
TRANSISTOR
CIRCUIT BOARD
SEMICONDUCTING MATERIALS HAVE SPECIAL PROPERTIES AND CAN BE USED AS AMPLIFIERS OR ELECTRICAL ON/OFF SWITCHES.
When each component of the circuit is separate, they are called discrete components. Each transistor and other parts of the circuit had to be soldered into place. Because of this, computers in the past were bulky and limited in how many transistors could be used.
THE INTEGRATED CIRCUIT WOULD FIX THIS!
THE INTERGRATED CIRCUIT (IC) —ALSO— KNOWN AS THE COMPUTER CHIP!
THE FIRST IC WAS INVENTED IN 1958 WHEN DISCRETE COMPONENTS OF A CIRCUIT WERE ETCHED ON A SINGLE PIECE OF SEMICONDUCTOR MATERIAL.
THE FIRST PLANAR IC WAS FABRICATED IN 1960.
MODERN COMPUTER CHIPS ARE MADE WITH A PROCESS CALLED PHOTOLITHOGRAPHY.
TODAY, ICs HAVE BILLIONS OF TRANSISTORS, SOME AS SMALL AS 2 NANOMETERS.
A STRAND OF DNA IS 2 NANOMETERS!
ULTRAVIOLET LIGHT
CHIP DESIGN
PHOTOMASK
LENS
TRANSISTORS KEEP GETTING SMALLER!
SILICON WAFER WITH PHOTORESIST COMPOUNDS

MEMORY AND STORAGE

COMPUTERS STORE AND ACCESS INFORMATION IN TWO MAIN WAYS

MEMORY

This stores data in the short term. RAM is the high-speed working memory of the computer; any file opened must be loaded into RAM for the CPU to access it. RAM is "volatile," which means data is temporary and will be lost when the computer is turned off. For example, an unsaved, newly written sentence in a text document is stored in RAM.

STORAGE

This stores data in the long term. When a computer is turned off, the data saved to storage remains; this means it is "nonvolatile." For example, a photo is saved to your hard drive using storage. There are different types of long-term memory, including ROM (read-only memory), which cannot be easily changed or written over. The programmed instructions a computer follows to boot up are stored in ROM.

Computers are constantly using different kinds of memory and storage. Volatile memory such as RAM is faster, smaller, and more expensive, while long-term storage, like on hard drives, is slower, larger, and cheaper.

As technology for memory and storage has advanced, their abilities have started to blend together. Speed, size, and cost are all taken into account when designing a computer. Many different memory and storage technologies are combined to make computers that work smoothly and are affordable.

UNITS OF DATA STORAGE

A BIT IS THE SMALLEST UNIT OF DATA.

Data is stored in a computer as 1s and 0s, also known as bits.

A BYTE (B) IS 8 BITS.

Also a collection of 1s and 0s that can represent one character such as an A, 5, or %.

A KILOBYTE (KB) IS ABOUT 1,000 BYTES.

This page of text is about 300 KB.

A MEGABYTE (MB) IS ABOUT 1,000,000 BYTES.

A one-minute audio recording is about 1 MB.

A GIGABYTE (GB) IS ABOUT 1,000,000,000 BYTES.

A 30-minute standard-definition movie is about 1 GB.

A TERABYTE (TB) IS ABOUT 1,000,000,000,000 BYTES.

A TB can hold more than 300,000 12-megapixel photos.

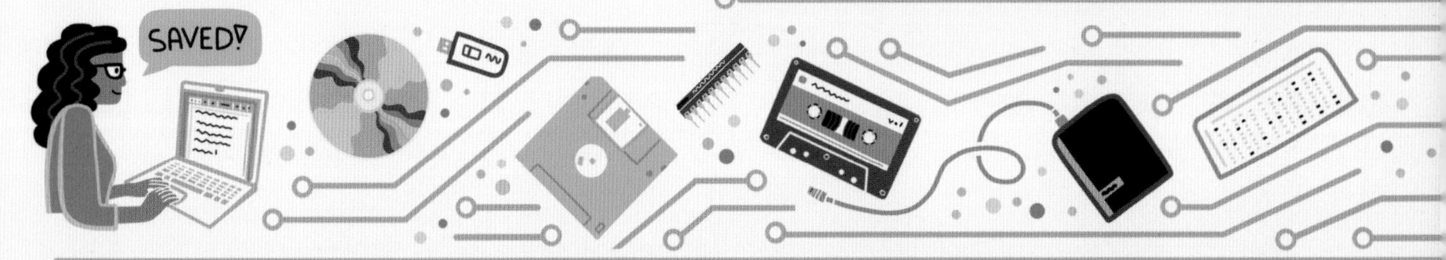

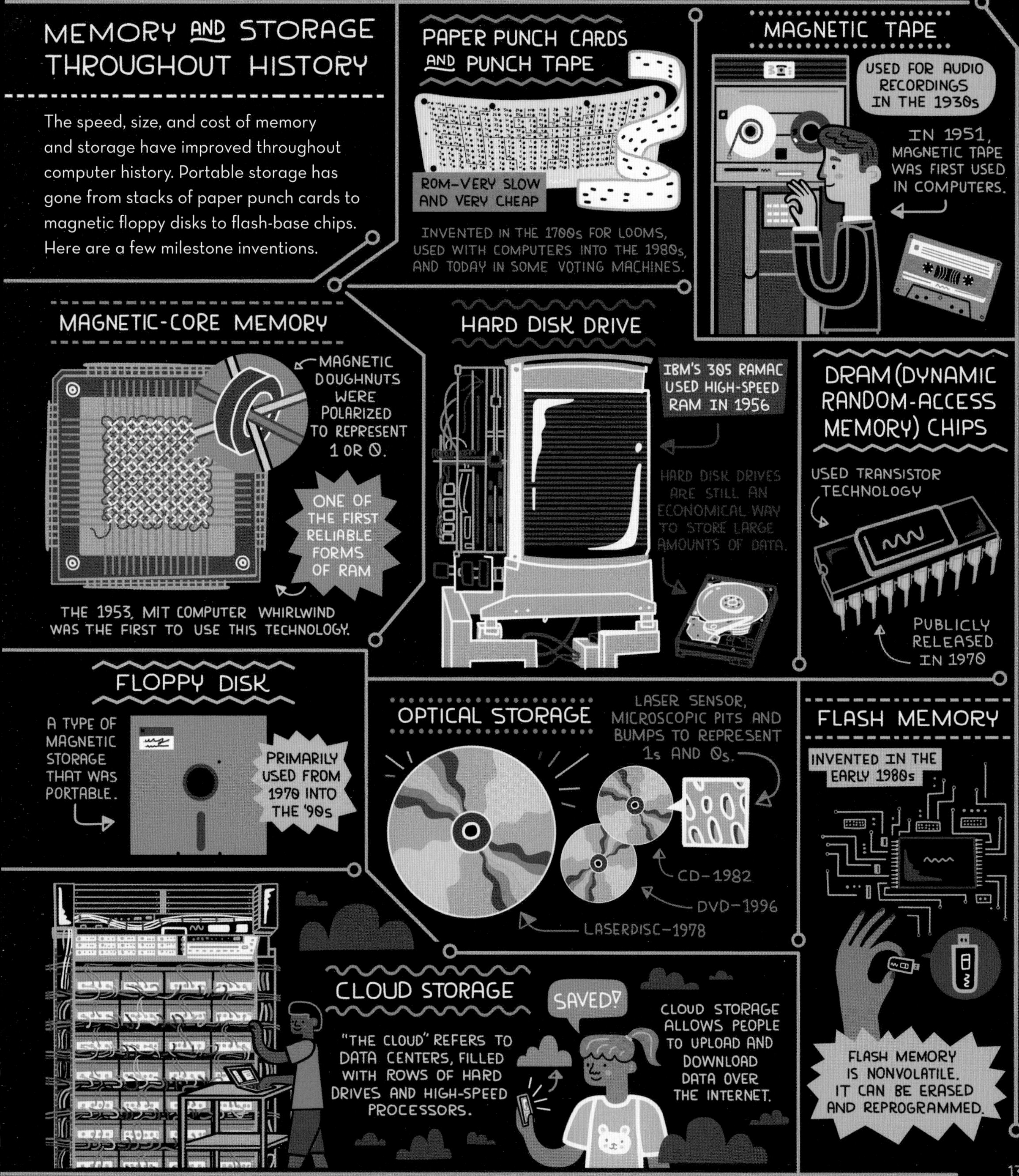

MEMORY AND STORAGE THROUGHOUT HISTORY
The speed, size, and cost of memory and storage have improved throughout computer history. Portable storage has gone from stacks of paper punch cards to magnetic floppy disks to flash-base chips. Here are a few milestone inventions.
PAPER PUNCH CARDS AND PUNCH TAPE
ROM—VERY SLOW AND VERY CHEAP
INVENTED IN THE 1700s FOR LOOMS, USED WITH COMPUTERS INTO THE 1980s, AND TODAY IN SOME VOTING MACHINES.
MAGNETIC TAPE
USED FOR AUDIO RECORDINGS IN THE 1930s
IN 1951, MAGNETIC TAPE WAS FIRST USED IN COMPUTERS.
MAGNETIC-CORE MEMORY
MAGNETIC DOUGHNUTS WERE POLARIZED TO REPRESENT 1 OR 0.
ONE OF THE FIRST RELIABLE FORMS OF RAM
THE 1953, MIT COMPUTER WHIRLWIND WAS THE FIRST TO USE THIS TECHNOLOGY.
HARD DISK DRIVE
IBM'S 305 RAMAC USED HIGH-SPEED RAM IN 1956
HARD DISK DRIVES ARE STILL AN ECONOMICAL WAY TO STORE LARGE AMOUNTS OF DATA.
DRAM (DYNAMIC RANDOM-ACCESS MEMORY) CHIPS
USED TRANSISTOR TECHNOLOGY
PUBLICLY RELEASED IN 1970
FLOPPY DISK
A TYPE OF MAGNETIC STORAGE THAT WAS PORTABLE.
PRIMARILY USED FROM 1970 INTO THE '90s
OPTICAL STORAGE
LASER SENSOR, MICROSCOPIC PITS AND BUMPS TO REPRESENT 1s AND 0s.
CD—1982
DVD—1996
LASERDISC—1978
FLASH MEMORY
INVENTED IN THE EARLY 1980s
FLASH MEMORY IS NONVOLATILE. IT CAN BE ERASED AND REPROGRAMMED.
CLOUD STORAGE
"THE CLOUD" REFERS TO DATA CENTERS, FILLED WITH ROWS OF HARD DRIVES AND HIGH-SPEED PROCESSORS.
SAVED!
CLOUD STORAGE ALLOWS PEOPLE TO UPLOAD AND DOWNLOAD DATA OVER THE INTERNET.

VIDEO GAMES

Beginning in the mid-twentieth century, the joy of video games inspired innovation in the field of computer science. From 1948 to 1951, the U.S. military developed a flight simulator called Whirlwind and created one of the very first on-screen graphic interfaces. What did they immediately do with this new technology? Made a video game, of course! It was a simple "ball" of light that would land in a moving "hole." Since then, video games have helped drive advancements in graphics, networking, and other computing technology.

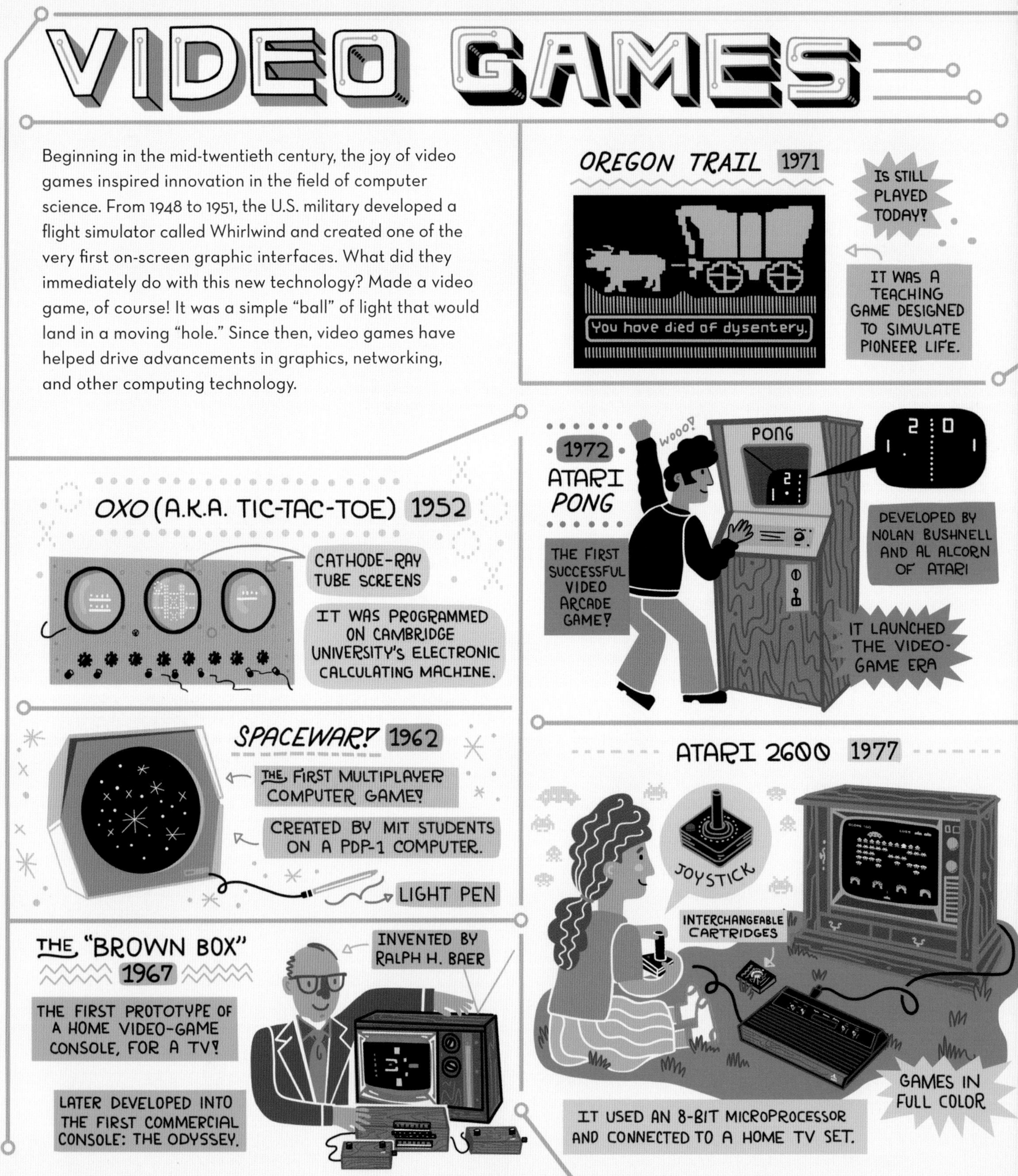

PAC-MAN
1980
MADE BY JAPANESE VIDEO-GAME DESIGNER TORU IWATANI.
BECAME THE BESTSELLING ARCADE GAME OF ALL TIME.
NES (NINTENDO ENTERTAINMENT SYSTEM) IS RELEASED IN THE U.S. 1985
NINTENDO REVITALIZED THE GAMING INDUSTRY IN THE U.S.
NINTENDO GAME BOY 1989
CARTRIDGE
ZELDA
ONE OF THE MOST POPULAR HANDHELD ELECTRONIC GAMES WITH REMOVABLE CARTRIDGES.
MORE THAN 100 MILLION GAME BOYS HAVE BEEN SOLD.
BEEP
BLOP
HALO
MICROSOFT RELEASES XBOX
XXXX 2001 XXXX
MICROSOFT CO-DESIGNED A GPU WITH NVIDIA TO ALLOW FOR AWESOME-LOOKING GAMES.
I'M PWNING THE NOOBS!
BY 2005, MILLIONS OF GAMERS USED THE XBOX FOR ITS ONLINE GAME PLAY AND HIGH-PREFORMING GRAPHICS.
WORLD OF WARCRAFT (WOW) GOES ONLINE
2004
AN ONLINE ROLE-PLAYING GAME THAT CONNECTED USERS ALL OVER THE GLOBE TO GO ON QUESTS TOGETHER.
PLAYED ON A PC
HI, GUILD! TIME TO RAID!
WORLD OF WARCRAFT
2006
NINTENDO Wii
USED GESTURE RECOGNITION AND REAL-TIME PHYSICAL MOVEMENT FROM THE PLAYER.
MOTION-SENSITIVE REMOTES
POKÉMON GO 2016
HELPED MAKE "AUGMENTED REALITY" MAINSTREAM.
GOTTA CATCH 'EM ALL!
USERS ALL OVER THE WORLD "CAUGHT POKÉMON" BY COMBINING GAME PLAY WITH THEIR PHONES.
MAINSTREAM VIRTUAL REALITY 2016
IVAN SUTHERLAND, A PIONEER IN COMPUTER GRAPHICS, MADE THE FIRST PROTOTYPE FOR A HEAD-MOUNTED VR (VIRTUAL REALITY) DISPLAY IN 1967.
BY 2016, HUNDREDS OF COMPANIES HAD RELEASED AFFORDABLE VR HEADSETS.

AI AND ROBOTS

WHAT IS AI?

AI (artificial intelligence) and machine learning are an entire branch of computer science. A computer learns based on an algorithm (step-by-step instructions) and by sifting through "training data." Once the computer has enough data, it creates a mathematical model to work with new, unlabeled data. Like a person learning something new, an AI needs practice. To be effective, an AI needs both a ton of data from which to learn and a very powerful computer to process it. Following are some exciting moments in AI history.

WHAT IS A ROBOT?

A *robot* is a machine that carries out a series of physical actions, guided by a computer or specialized program. Robots have automated certain kinds of factory work, performing tasks that are tedious or dangerous for workers. Robots and automation have made goods less expensive over time but have also caused the loss of many traditional jobs, similar to what happened during the Industrial Revolution. Some robots are built with decision-making capability by using AI, and some follow a set of simple instructions. Here are a few milestones in robotics history.

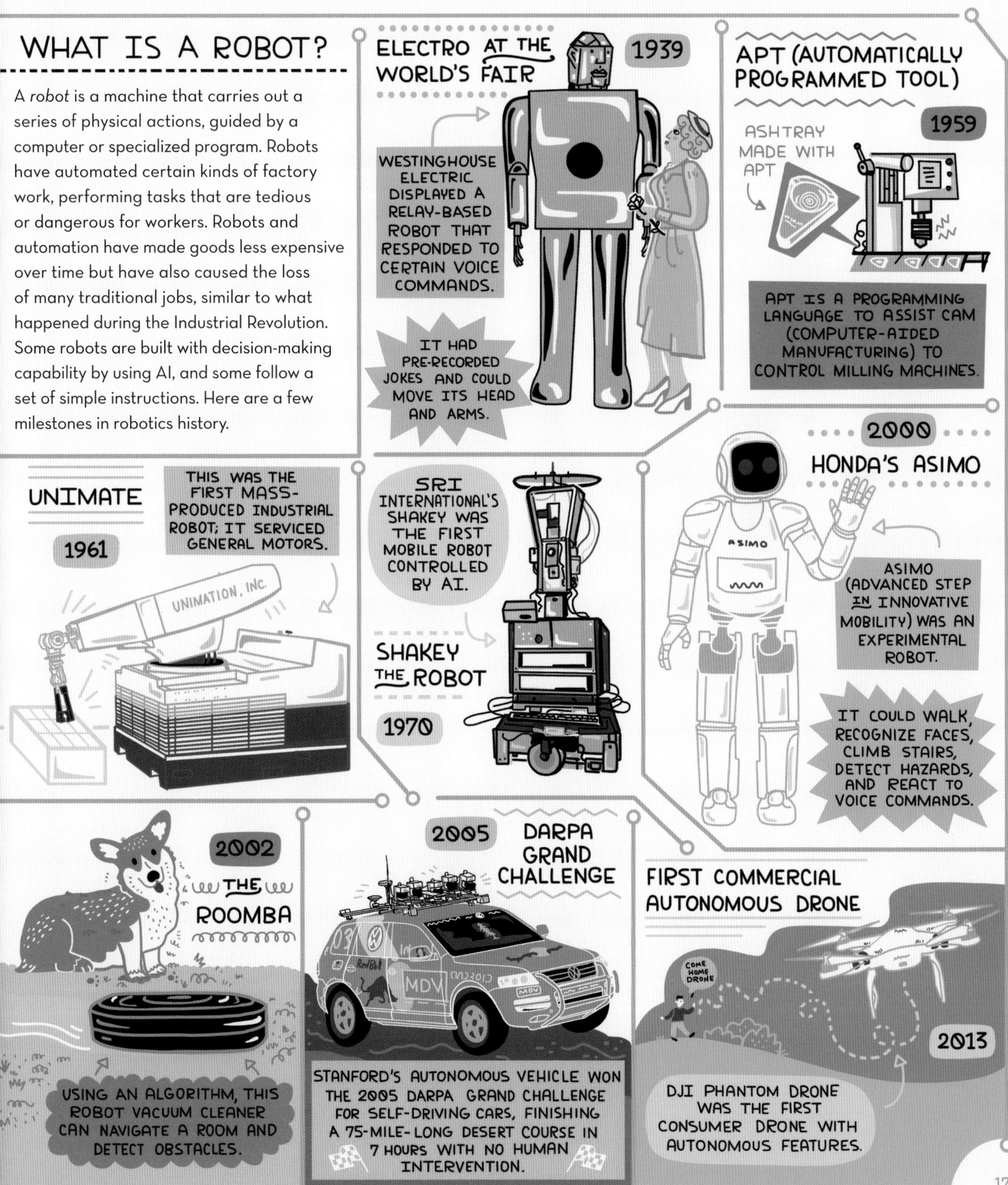

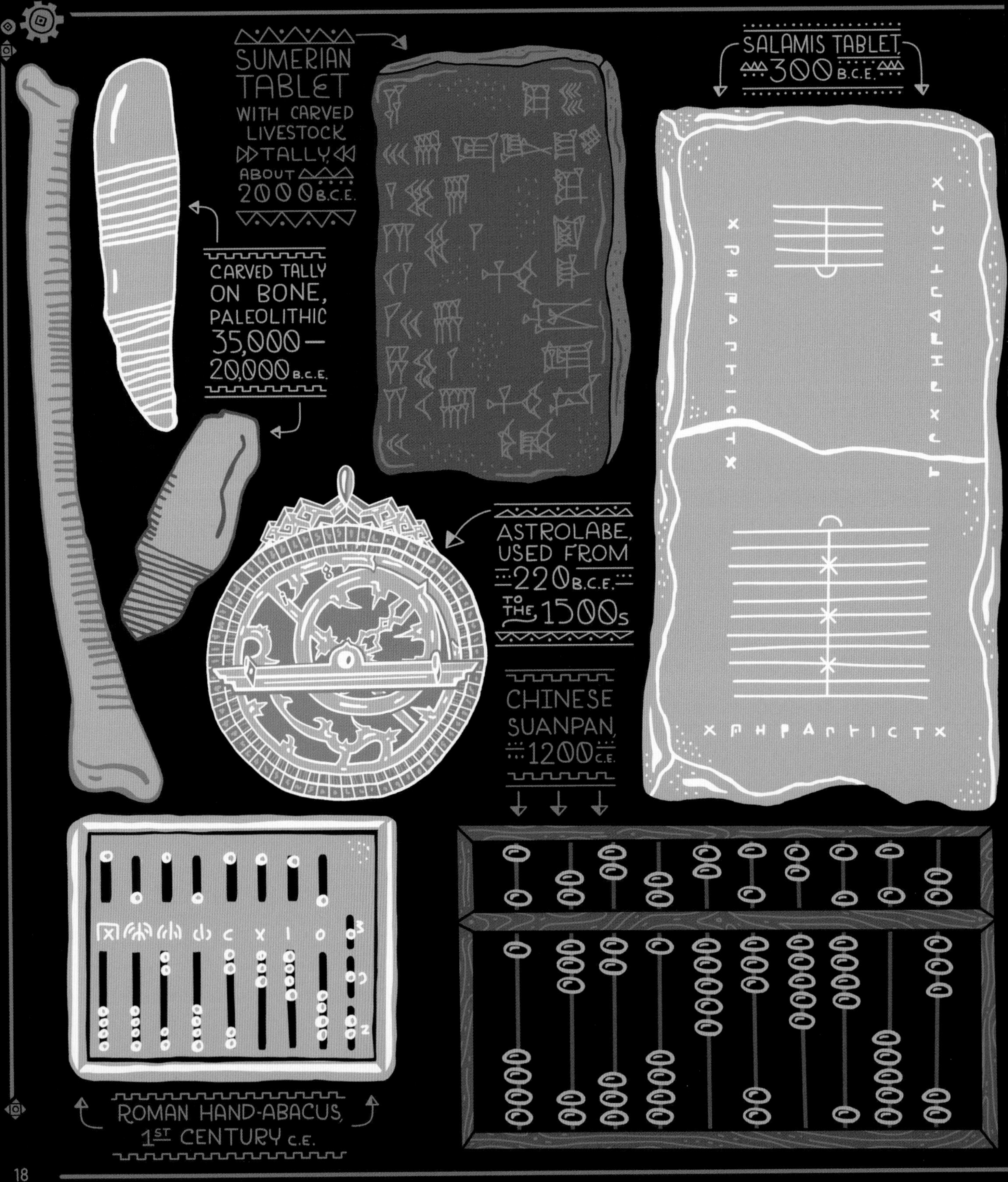
SUMERIAN TABLET WITH CARVED LIVESTOCK TALLY, ABOUT 2000 B.C.E.
SALAMIS TABLET, 300 B.C.E.
CARVED TALLY ON BONE, PALEOLITHIC 35,000–20,000 B.C.E.
ASTROLABE, USED FROM 220 B.C.E. TO THE 1500s
CHINESE SUANPAN, 1200 C.E.
ROMAN HAND-ABACUS, 1ST CENTURY C.E.

ANCIENT CIVILIZATIONS

25,000 B.C.E.–1599 C.E.

COUNTING AND CALCULATING

Let's start at the very beginning, long before the first electronic computer was ever plugged in. People needed to do simple math. Questions such as "How many babies are being born?" and "Are there enough sheep in the flock?" were simple enough to keep track of with a tally; but as society grew, so did the need for complex calculation. Ancient empires created marvels, like the Mayan pyramids, the Great Sphinx of Giza, and the Roman Colosseum. These wonders, and the civilizations that built them, relied on the ability to record data and to tabulate quantities that were much larger than anyone could keep track of in their head.

Around the world, different tools, like counting boards and abacuses, were invented so people could do calculations beyond their own mental capacity. Along with such tools, new ways of keeping numerical records were developed, and devices were built to chart the stars and aid in the task of telling time. These new technologies were used by everyone from farmers and merchants to the bureaucrats who ran the governments of empires. As trade and enterprise grew, so did the study of mathematics. The polymaths and inventors of the ancient world even dreamed of creating robots that would move or play music all by themselves!

Although the distant past may seem very different from today, in both, technology enhanced one's thinking power and allowed people to build and dream bigger than ever before.

TIMELINE

THIS BABOON FIBULA WITH 29 NOTCHES WAS FOUND IN THE MOUNTAINS OF SWAZILAND.

ABOUT 35,000 B.C.E.

THE LEBOMBO BONE

Archaeologists have found animal bones with tally notches carved into them. This was one of the ways that prehistoric humans kept number records. The Lebombo bone is one of the oldest known mathematical artifacts.

300 B.C.E.

BABYLONIAN PLACEHOLDER FOR 0

The Babylonians started using two slanted wedges to represent an empty space on an abacus. It was used as a placeholder. It did not represent zero as a numerical value, but instead was used as punctuation.

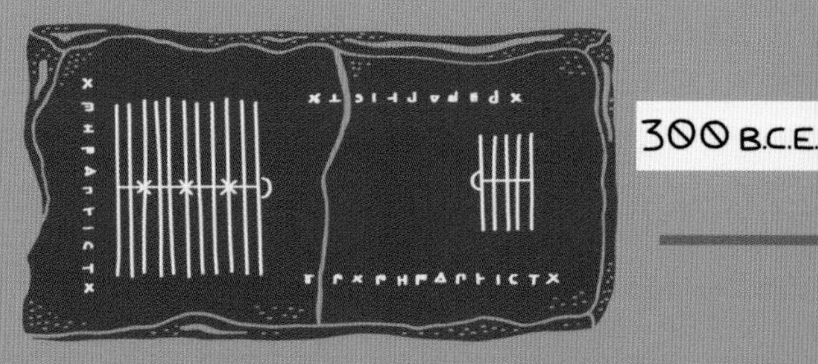

300 B.C.E.

THE SALAMIS TABLET

Discovered in Greece, this counting board is one of the oldest surviving calculating devices and a precursor to the modern abacus.

1 2 3 4 5 6 7 8 9 0

HINDU-ARABIC NUMERALS

Our modern positional decimal number system was developed in India. Hindu-Arabic numerals include symbols to represent the ten digits from 0 to 9. This leap in mathematics created a new, quick method of doing arithmetic with ink and paper instead of using an abacus.

2500 B.C.E.

THE SUMERIAN ABACUS

Historians believe that the Sumerians of Mesopotamia invented the first abacus. It was most likely a flat stone notched with parallel lines where counters, like pebbles, were placed to indicate value.

475 B.C.E.

CHINESE ROD NUMERALS

As early as the Warring States Period in China, counting rods were used by merchants, astronomers, and state officials. They allowed users to do addition, subtraction, multiplication, and division with efficiency and speed.

150 B.C.E.

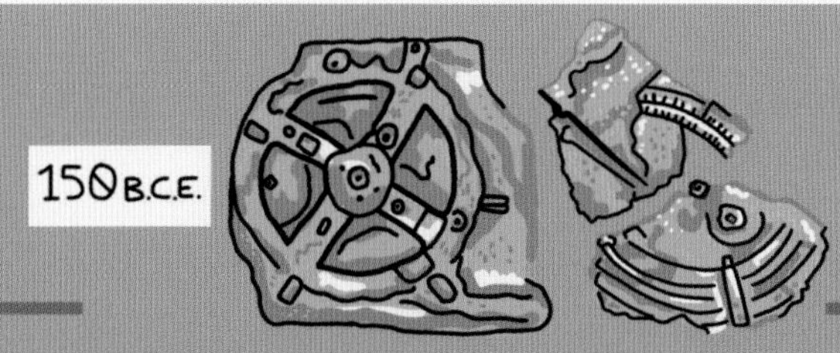

THE ANTIKYTHERA MECHANISM

The most complicated mechanical device discovered from this time period, the Antikythera mechanism was used to calculate astronomical events in ancient Greece. Although it was just a series of gears, many historians have nicknamed it "the world's first computer."

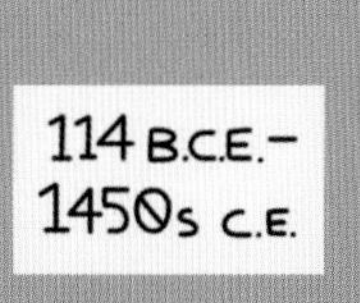

114 B.C.E.–1450s C.E.

THE SILK ROAD

Instrumental in the development of science and mathematics in the ancient world by allowing the exchange of ideas and philosophy as well as goods, the Silk Road was a network of land and sea trade routes between Europe, the Middle East, South and East Asia, and East Africa.

683 C.E.

THE FIRST-KNOWN 0

K-127 is one of the earliest surviving artifacts that shows 0 being used. Discovered in Cambodia, the stone stele has an inscription in Old Khmer, reading "The Chaka era reached the year 605 on the fifth day of the waning moon."

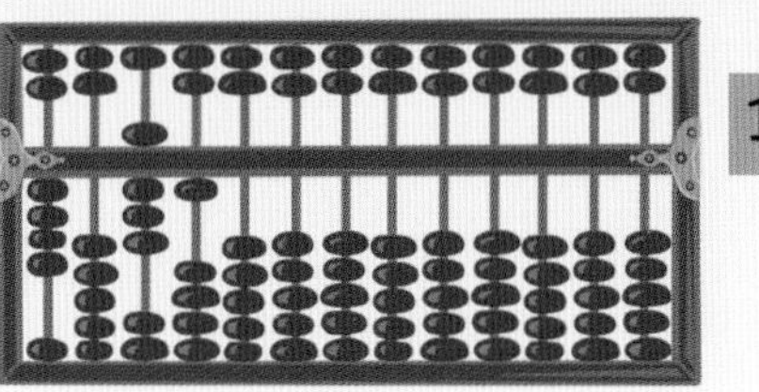

1200 C.E.

THE SUANPAN

The earliest writing about the Chinese abacus dates to circa 190 C.E. The modern abacus, called the suanpan, features a "5 plus 2" design and is thought to have been developed around 1200 C.E. It is still used today across the world.

STORIES FROM HISTORY

The history of the computer starts at the dawn of civilization, when prehistoric people began to count. A very long time ago, our ancestors grouped things in only three ways: one, two, and many. People needed to know exactly how many items were in a group and began counting, using their fingers (and sometimes even their toes).

As small tribes became larger communities, using ten fingers to count wasn't enough. People kept count by painting on rocks, collecting pebbles, tying knots, or carving notches into sticks or animal bones. These were just some of the different ways that prehistoric tribes kept tallies. Historians speculate that early people documented everything from the number of goats in a herd to people in a tribe.

THE WORD *ABACUS* COMES FROM THE GREEK WORD FOR "SLAB" (ABAX).

THE FIRST TABULATING TOOLS

As communities grew into cities and empires, the need to calculate and keep data records also increased. Merchants needed to keep track of the goods they were selling. Military officials needed to calculate the number of people who could fight as soldiers. Governments needed to know how much food should be grown and how much tax to collect. City planners and early engineers needed to calculate the lengths of aqueducts and details of other infrastructure projects. Tools were built to aid in these calculations.

It is important to understand that what is known about ancient and medieval history is very limited. Our knowledge of the past is based on the artifacts that have been preserved and studied. Much of the truly ancient past has been lost, including knowledge that was passed down orally, inventions made from biodegradable materials (like wood) and items and records that were destroyed by invading forces and colonizers.

Historians believe that the first known tool made specifically to tabulate numbers, the abacus, was invented by the Sumerians in Mesopotamia around 2500 B.C.E. The Sumerian abacus did not look like a modern abacus on rods. It was a counting board made of wood, clay, or stone with parallel etched lines in which the user would place pebbles or sticks to show value. It's believed that before this counting board, people would simply draw a counting table in the sand or dirt. The Sumerian abacus was sturdier and more organized than a drawing on the ground. As counting boards and abacuses were developed around the world, they were used to add, subtract, multiply, and divide large sums with efficiency and speed.

In ancient Rome, merchants, engineers, and tax collectors carried portable abacuses for their work. Archaeologists have found a Roman hand-abacus that dates back to the first century C.E. In ancient China, instead of abacuses, people carried around bags of counting rods made of ivory, bamboo, or steel. Counting rods were used from around 475 B.C.E. well into the 1500s C.E.

OUR NUMBER SYSTEM IS BASED IN 10s BECAUSE EARLY HUMANS COUNTED ON THEIR FINGERS.

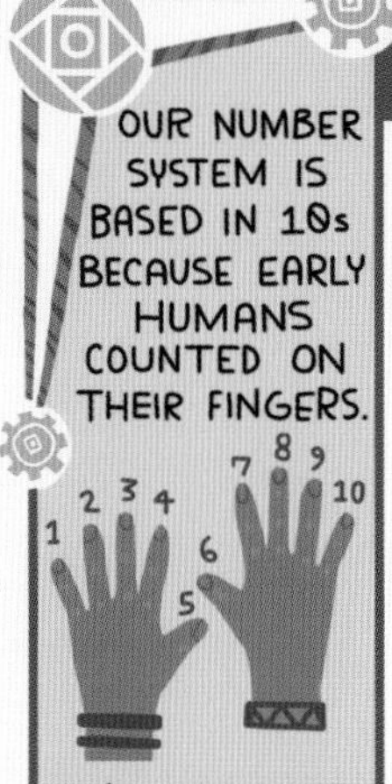

IT'S WHY WE CALL NUMBERS *DIGITS*!

THE SUMERIAN ABACUS WAS BASED IN A SEXAGESIMAL (60) SYSTEM.

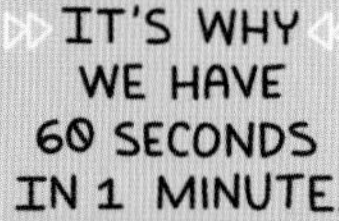

IT'S WHY WE HAVE 60 SECONDS IN 1 MINUTE.

THE ASTROLABE WAS AN ANALOG CALCULATOR USED FROM ANCIENT GREECE THROUGH THE 6TH CENTURY TO HELP SAILORS NAVIGATE BY THE POSITIONS OF STARS.

NUMBERS IN THE ANCIENT WORLD

AZTEC	SYMBOL							
	VALUE	1	20	400	8,000			
SUMERIAN	SYMBOL							
	VALUE	1	10	60	600	3,600	36K	216K
ROMAN	SYMBOL	I	V	X	L	C	D	M
	VALUE	1	5	10	50	100	500	1,000
EGYPTIAN	SYMBOL							
	VALUE	1	10	100	1,000	10K	100K	1 MIL

THE DECIMAL NUMERAL SYSTEM

A huge leap forward in mathematics happened in India during the sixth or seventh century C.E. when Hindu-Arabic numerals were created. Unlike number systems of the past, in this system each numeral represented a digit from 0 to 9 and was designed for positional notation in a decimal system. This allowed people to do math quickly on paper. It opened the door for algebra, logarithms, and modern mathematics. As trade grew across Eurasia, so did the exchange of mathematical ideas and technology. Hindu-Arabic numerals became widespread in the twelfth century and were especially embraced in the Middle East; that is why they are commonly known as Arabic numerals today.

ANCIENT AUTOMATONS

Many engineers, philosophers, and polymaths of the ancient world dreamed of creating automatons and programmable machines. In 60 C.E., Hero (also known as Heron) of Alexandria wrote about his designs for a robotic contraption that involved a cart moved by "programing" a system of strings, pulleys, and weights. Hero is also credited with inventing the first vending machine. It was a coin-operated device that used levers to dispense holy water.

In the Middle East, inventors working at the House of Wisdom (known as the Grand Library of Baghdad) also dreamed of robots. In 1206 C.E., polymath Ismail al-Jazari wrote *The Book of Knowledge of Ingenious Mechanical Devices*, in which he describes many machines, including a music box–like automaton band that was placed on a small boat and entertained guests at royal parties.

ONE OF AL-JAZARI'S FAMOUS AUTOMATONS WAS AN ELEPHANT CLOCK (1206 C.E.)

IMPACT OF THE ERA

Technology and inventions, like the abacus and Hindu-Arabic numerals, allowed humanity to collectively leap forward. This theme is repeated throughout history—technology is built out of necessity and allows humanity as a whole to tackle more and more complicated problems.

SCYTALE — 700 B.C.E.

Just as modern-day militaries depend on secrecy when sending messages, ancient empires also needed to send reports that could not be read by the enemy if intercepted. The Spartan military of ancient Greece developed a tool, called the scytale, to keep communication coded.

A scytale was made up of two identical wooden staffs, one for the sender and one for the recipient. Whenever information needed to be sent, the sender wrote it on a scroll of parchment wrapped around one staff. When the message was ready for transport, it was unwrapped from the staff and became scrambled and unreadable. Only when the recipient rewrapped the parchment on their own staff did the words make sense again. This is one of the earliest known tools used to create coded transmissions and is one of the first cryptographic devices.

EXAMPLES OF CRYPTOGRAPHY HAVE BEEN FOUND IN ANCIENT EGYPT, MESOPOTAMIA, AND JUDEA, BUT THE SCYTALE IS ONE OF THE FIRST-KNOWN DEVICES SPECIFICALLY MADE TO ENCRYPT MESSAGES.

THE DIAMETER OF THE STAFF ACTED AS AN ENCRYPTION KEY.

QUIPU — 1400-1532 C.E.

As the massive Inca Empire expanded across the Andean highlands, it had an estimated population of twelve million people; this meant there was a great need to collect and record data. The quipu was an incredibly complex accounting device made of multicolored cotton cords that were knotted in various ways to record data. Historians believe that quipus kept statistics and census data, recorded events, were utilized as calendars, and used to send messages.

Quipus have been well preserved because of the dry climate of the Andes mountains. Like many aspects of pre-Columbian civilizations, there is still much we don't know about the quipu. We do know that it was an ingenious way to keep historical records based on math and that it was used at every level in Incan bureaucracy.

THE TYPE OF KNOT AND ITS PLACEMENT, GROUPING, AND COLOR EACH HAD SPECIAL NUMERICAL MEANING.

THE ANTIKYTHERA MECHANISM — 150 B.C.E.

In 1901, sponge divers off the coast of Greek island Antikythera discovered a shipwreck filled with statues and artifacts from 150 B.C.E. Among the treasures was a slimy, corroded, green piece of metal, now called the Antikythera mechanism. In the 1970s and '90s, historians used X-ray imaging to reveal that it was not just a hunk of junk but a piece of ancient technology. In 2006, scholars used computed tomography imaging that revealed inscriptions and complex gears. The Antikythera mechanism is the most sophisticated device known from the ancient world, and it wouldn't be matched for the next thousand years.

Ancient Greeks operated the device by turning the mechanism's hand crank, and the many interlocking bronze gears inside would predict astrological events, the phases of the moon, eclipses, calendar cycles, solstices, and the dates of the Olympics. Historians speculate that it could have been used to plan crop planting, for religious astrology purposes, in scientific study, and for military strategy.

THE ABACUS AROUND THE WORLD

FUNDAMENTALS OF AN ABACUS

For centuries, counting boards and tools that resembled the modern abacus were used to do calculations quickly. Although many cultures arranged their abacuses in different ways, the principles were usually the same. Generally, each rod represented a place value, and each bead represented a numeric value. By sliding the beads up and down (or left and right), a person could use an abacus to keep track of sums, carries, and other important numbers.

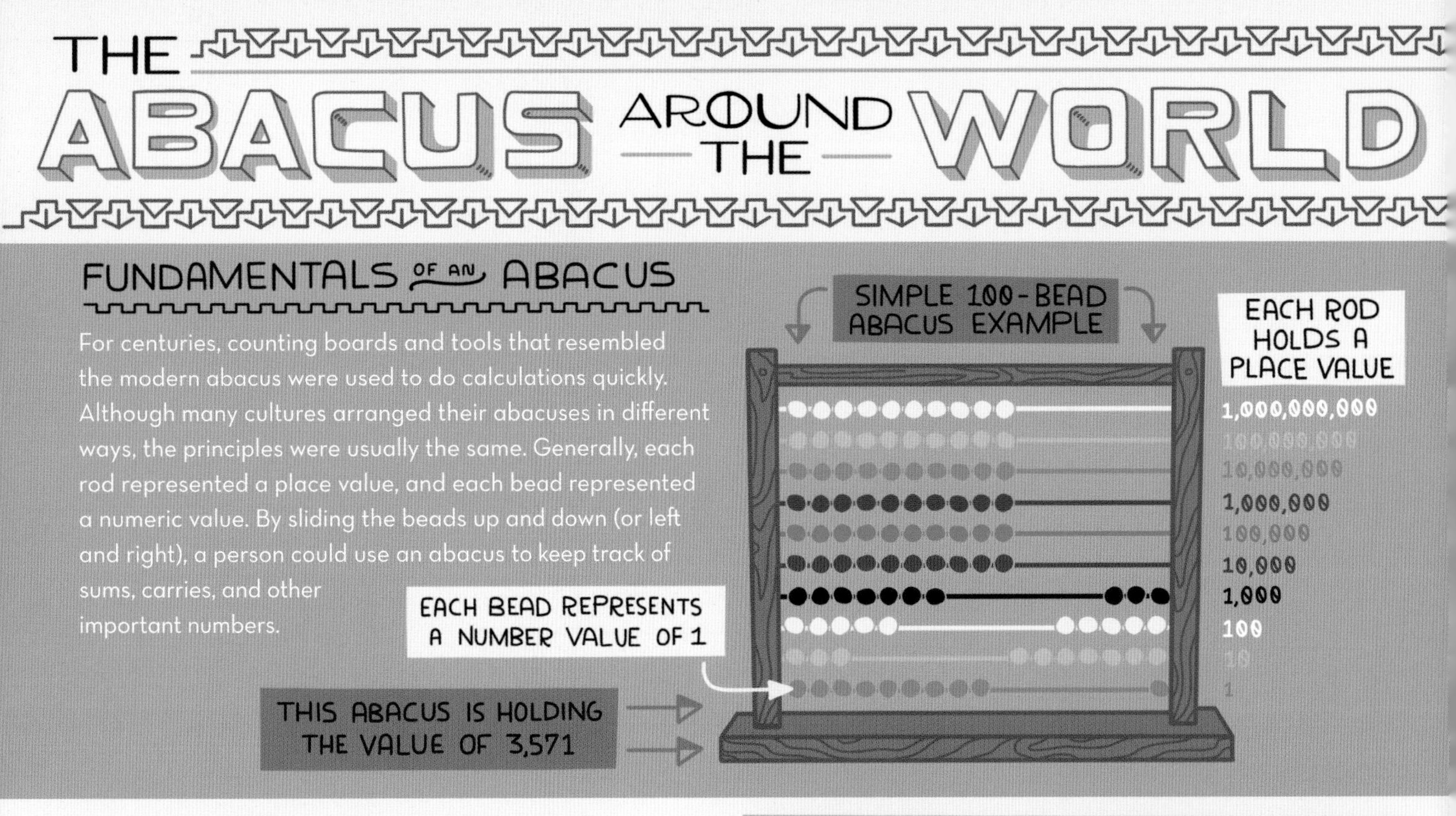

THE ROMAN HAND-ABACUS—1 C.E.

One of the earliest-known portable counting boards used by merchants, bureaucrats, and engineers in ancient Rome.

THE ROMAN HAND-ABACUS WAS BASED ON 12 BECAUSE THE ROMAN POUND WAS DIVIDED INTO 12 OUNCES.

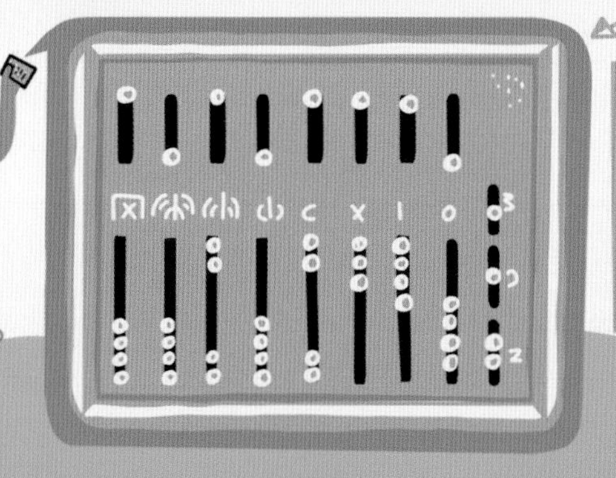

MADE OF A METAL PLATE AND BEADS THAT WOULD BE MOVED IN SLOTS.

NEPOHUALTZINTZIN—900–1000 C.E.

This is a Mesoamerican abacus with 13 rows and 7 beads. Multiples of the Nepohualtzintzin's 91 beads were used to represent the seasons in the year, the number of days in a corn harvest cycle, the days in a pregnancy, and the days in a calendar year. Etchings and writings are all that's left of the Nepohualtzintzin due to the destruction of pre-Columbian artifacts after Spanish colonization.

ANCIENT MAYAN ARTIFACTS DEPICTING THE NEPOHUALTZINTZIN HAVE BEEN FOUND.

THERE IS EVIDENCE THAT THIS ABACUS WAS A WEARABLE BRACELET.

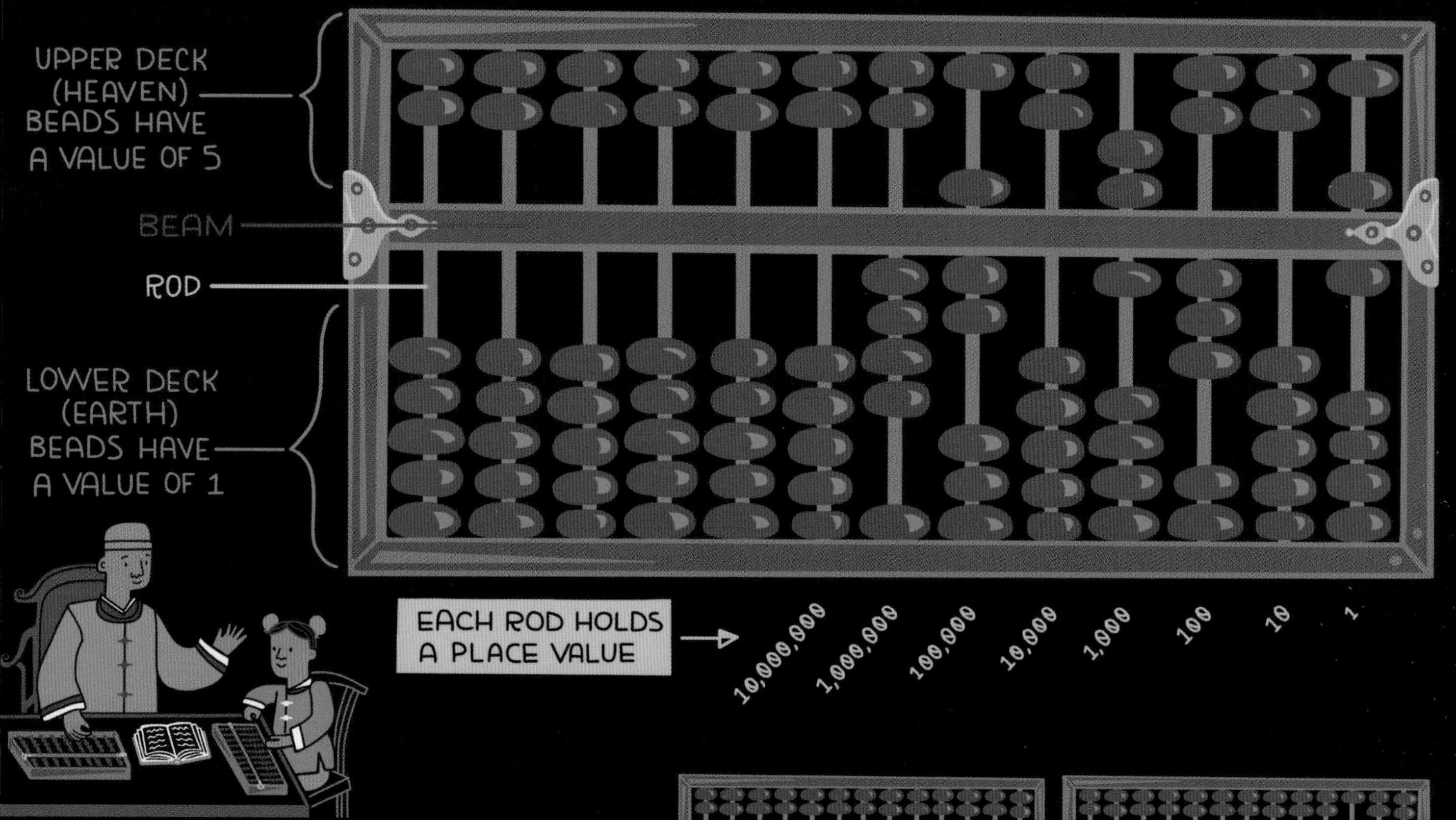

SUANPAN—1200 C.E.

Historians are unsure of exactly when the abacus was first used in China. In the year 1200 C.E., the modern Chinese abacus, the "5 plus 2" suanpan, was invented. With a skilled hand, certain calculations on the suanpan could be done as fast as on a digital calculator.

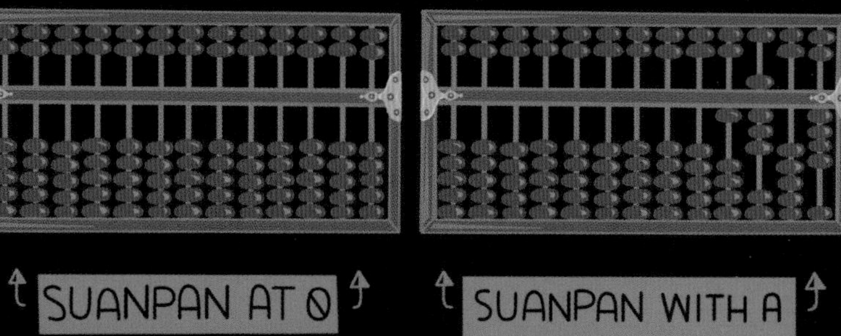

THE RUSSIAN TCHOTY

THE JAPANESE SOROBAN

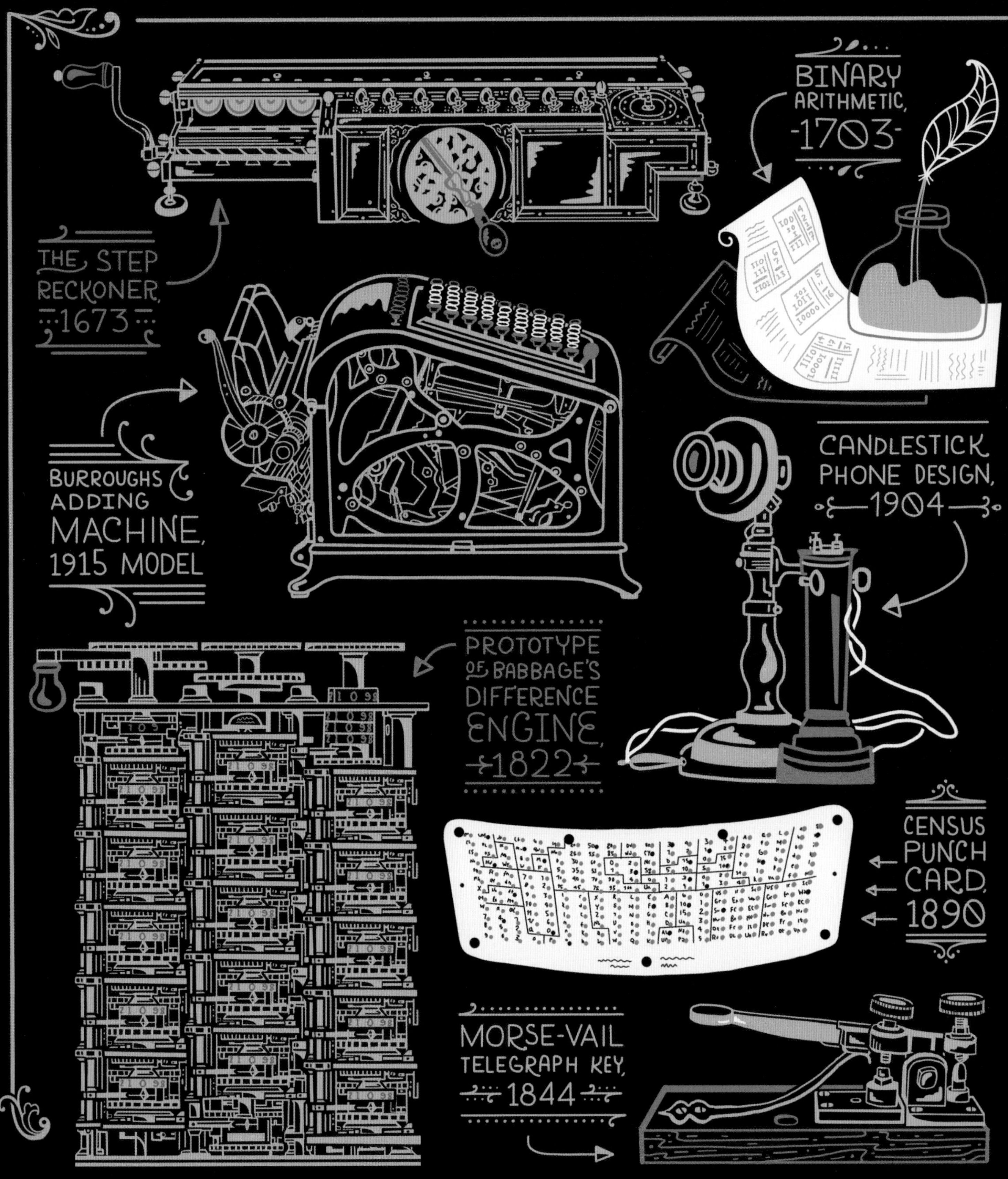

BINARY ARITHMETIC, 1703
THE STEP RECKONER, 1673
BURROUGHS ADDING MACHINE, 1915 MODEL
CANDLESTICK PHONE DESIGN, 1904
PROTOTYPE OF BABBAGE'S DIFFERENCE ENGINE, 1822
CENSUS PUNCH CARD, 1890
MORSE-VAIL TELEGRAPH KEY, 1844

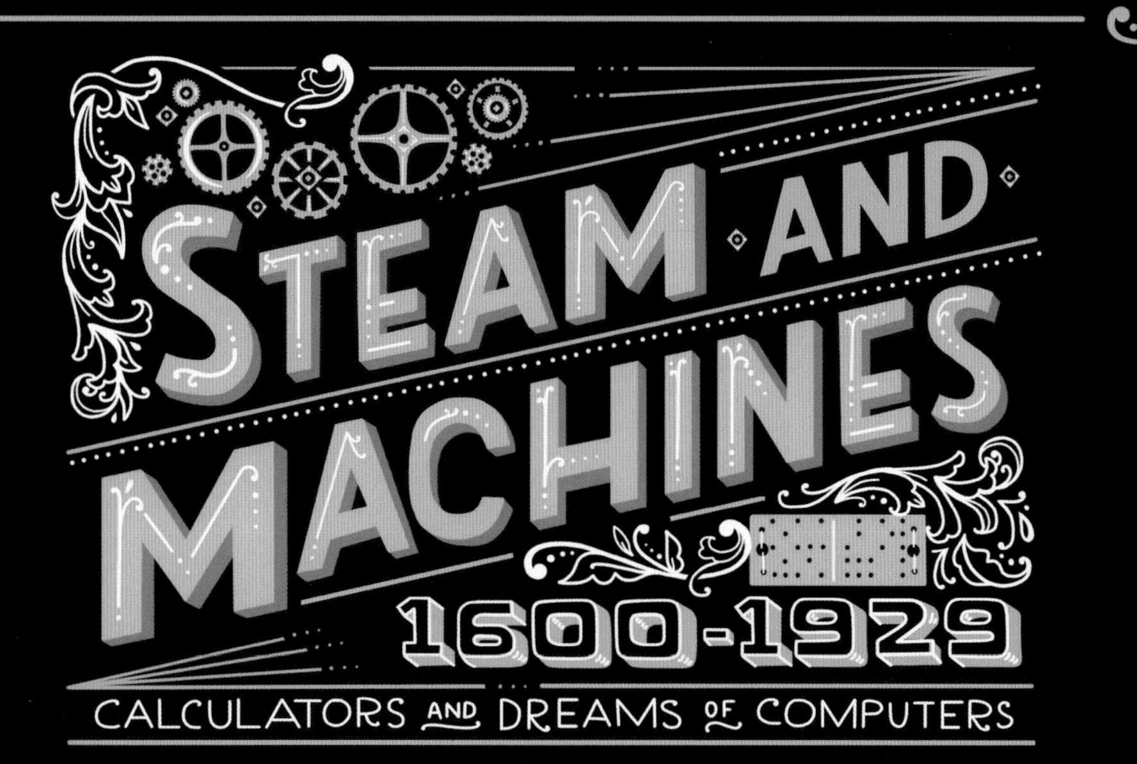

CALCULATORS AND DREAMS OF COMPUTERS

The Industrial Revolution transformed our world with automation. Although ancient Greeks had studied the potential of superheated water vapor, a practical steam engine was not built until the 1700s, when several engineering breakthroughs in iron manufacturing made it possible. A steam engine uses the energy of boiling water to spin a crank. The automation of this simple motion allowed inventors to build machines that transformed an engine's rotational force into tasks such as sewing fabric, pumping groundwater, and sawing lumber. By 1820, steam engines powered trains, ships, and factories. In these factories, the labor to make products was reorganized in a new way called an assembly line. Assembly lines broke down manual labor into specific repetitive tasks for each worker to complete faster (and cheaper) than ever before. Assembly lines and steam-powered tools allowed manufacturing on a massive scale.

Mathematics also advanced during this era. Human "computers" worked together to create complicated mathematical tables for easy reference. Similar to how manual labor was divided by assembly lines, human computers divided mental labor into small tasks to tackle big problems. Machines were invented that could aid in this repetitive mental work. It wasn't long before mathematicians dreamed of machines that could "think" for themselves.

In this new mechanical world, the United States dominated the market for business machines. Mathematical discoveries of this era, like binary arithmetic and Boolean algebra, would be applied to electrical devices in the future. In fact, many discoveries of this time would define computer history, even though the full potential of those ideas and devices was not yet realized. The Industrial Revolution was the primordial ooze in which the twentieth-century computer would be built.

TIMELINE

1613

FIRST USE OF THE WORD *COMPUTER*

The first printed use of the word *computer* was in poet Richard Brathwaite's book *The Yong Mans Gleanings*. It was not used in reference to a machine, however, but to a person who did math calculations as their job.

1621

INVENTION OF THE SLIDE RULE

The slide rule, invented by William Oughtred, is a portable mechanical device comprising two logarithmic scales. It was used into the 1970s by engineers for making specific calculations.

1760

THE INDUSTRIAL REVOLUTION BEGINS

Advances in technology and manufacturing created jobs in factories. This led to people moving away from farm life and into cities. Although the Industrial Revolution began in England, inventions such as electric generators and coal-powered steam engines would transform the entire world.

1834

THE ANALYTICAL ENGINE

The Analytical Engine was Charles Babbage's dream of a thinking, programmable machine, and it is considered the first working design for a general-purpose computer. Although a working Analytical Engine was never built, its design included many features of the modern computer.

1874

THE SEMICONDUCTOR DIODE IS INVENTED

In 1874, Karl Ferdinand Braun discovered that when a galena crystal is probed with a thin metal wire, it could conduct electrical current in a single direction. This is one of the properties that makes semiconductors useful in electronics.

1876

FIRST PHONE CALL

Alexander Graham Bell made the first telephone call, to his assistant, showing the world that his invention would let people "talk with electricity." By the 1920s, about a third of all U.S. households had a telephone.

HUMAN COMPUTERS PREDICT HALLEY'S COMET

Three French mathematicians worked together to chart the path of Halley's Comet. Each of them worked on different parts of the complicated math calculations, and this teamwork was a huge success! It led to many government-funded projects, including the hiring of large groups of human computers to create different kinds of mathematical tables.

GEORGE BOOLE

STATEMENT-A	STATEMENT-B	STATEMENT-A&B
TRUE	TRUE	TRUE
TRUE	FALSE	FALSE
FALSE	TRUE	FALSE
FALSE	FALSE	FALSE

1854 BOOLEAN ALGEBRA

George Boole published *An Investigation of the Laws of Thought*, in which he described the rules and reasoning for Boolean algebra. In 1936, engineer Claude Shannon realized that Boolean algebra could be used to describe the logic gates that would make up a computer's circuitry.

FIRST SPAM MESSAGE VIA TELEGRAPH

Infrastructures of wires were built to connect the growing economy of the nineteenth century, and people began communicating by electric telegraph in the early 1840s. In 1864, the first spam message was sent via telegraph—an advertisement for a group of dentists.

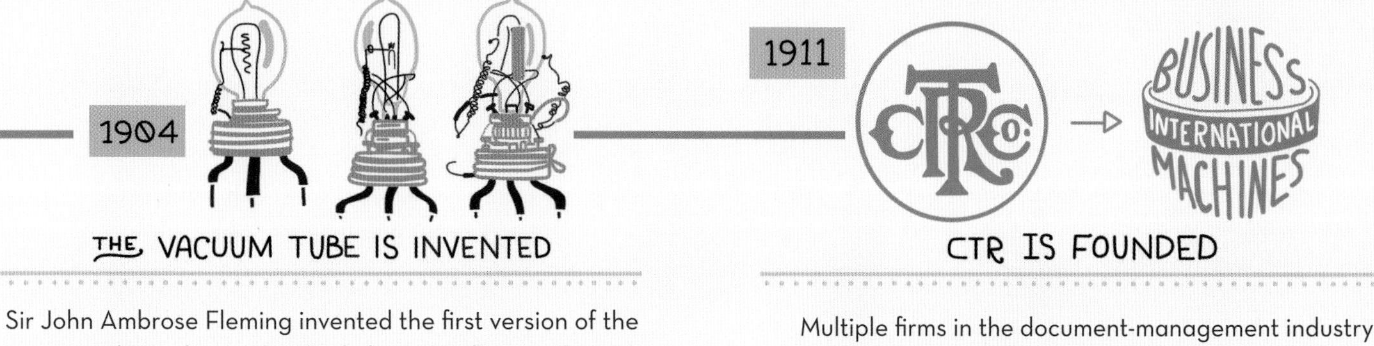

THE VACUUM TUBE IS INVENTED

Sir John Ambrose Fleming invented the first version of the vacuum tube, a device that had electricity flowing through it in a single direction. It was used as an amplifier in radios and televisions; decades later, a modified version was used in computers.

CTR IS FOUNDED

Multiple firms in the document-management industry merged to form CTR (Computing-Tabulating-Recording), which was renamed IBM (International Business Machines) in 1924.

STORIES FROM HISTORY

During the Industrial Revolution, new technology transformed the way that people did their jobs. Work that previously would have taken a skilled shoemaker or carpenter hours to accomplish could now be done in minutes or seconds with the help of new tools and a reorganization of labor into assembly lines. Goods were mass-produced, and traditional physical labor, like plowing a field, was replaced by new steam-powered inventions. This increased mechanical productivity put many laborers and farmers out of business, forcing them to leave rural communities to work in new urban factories. While this new industry boomed, it also meant low pay and harsh conditions for early factory workers. With steam-powered vehicles, global trade was no longer at the mercy of wind patterns and sailing ships. By the late nineteenth century, steamboats allowed mass-produced consumer goods to reach even the most remote settlements. This led to a massive increase in monetary wealth that, in turn, supercharged the demand for accurate accounting and quick mathematical calculations. Instead of tediously recalculating the same math problems over and over again, workers used printed mathematical tables. These pamphlets and books were created by groups of human computers, who worked together using specific algorithms. Different numerical tables were printed for all sorts of industries, from star charts for sailors to trigonometry tables for engineers. Anyone that needed to do math as part of their job relied on the tables created by a team of human computers.

THE DIFFERENCE ENGINE

Hot-shot gentleman mathematician Charles Babbage was tasked with checking mathematical tables for errors. He worked with astronomer John Herschel on star tables for the *Nautical Almanac*. Babbage hated doing this repetitive and boring task, and in frustration, he exclaimed, "I wish to God these calculations had been executed by steam!"

Human error and printing mistakes in mathematical tables were a huge source of anxiety to the government. The military depended on ballistics tables, just as ships depended on astronomical charts. Babbage saw an opportunity and designed a machine that could calculate polynomial tables and then print the table—all with the certainty of a grandfather clock! He called this mechanical calculator the Difference Engine. The British government was excited by this project and granted him £17,500 to build the machine. (At the time, this was enough money to buy two brand-new train engines!)

In 1822, Babbage built a small section of the Difference Engine as proof of concept. He also used it to dazzle guests at dinner parties. But alas, for

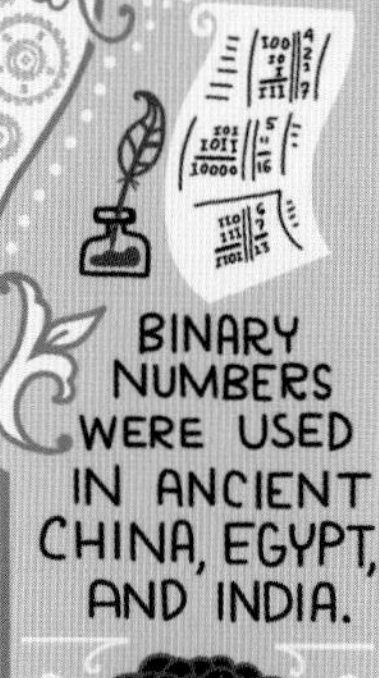

BINARY NUMBERS WERE USED IN ANCIENT CHINA, EGYPT, AND INDIA.

IN GERMANY, IN 1703, GOTTFRIED WILHELM LEIBNIZ CREATED RULES FOR BINARY ARITHMETIC.

IN 1838, MORSE CODE WAS DEVELOPED USING THE NEW ELECTRIC TELEGRAPH.

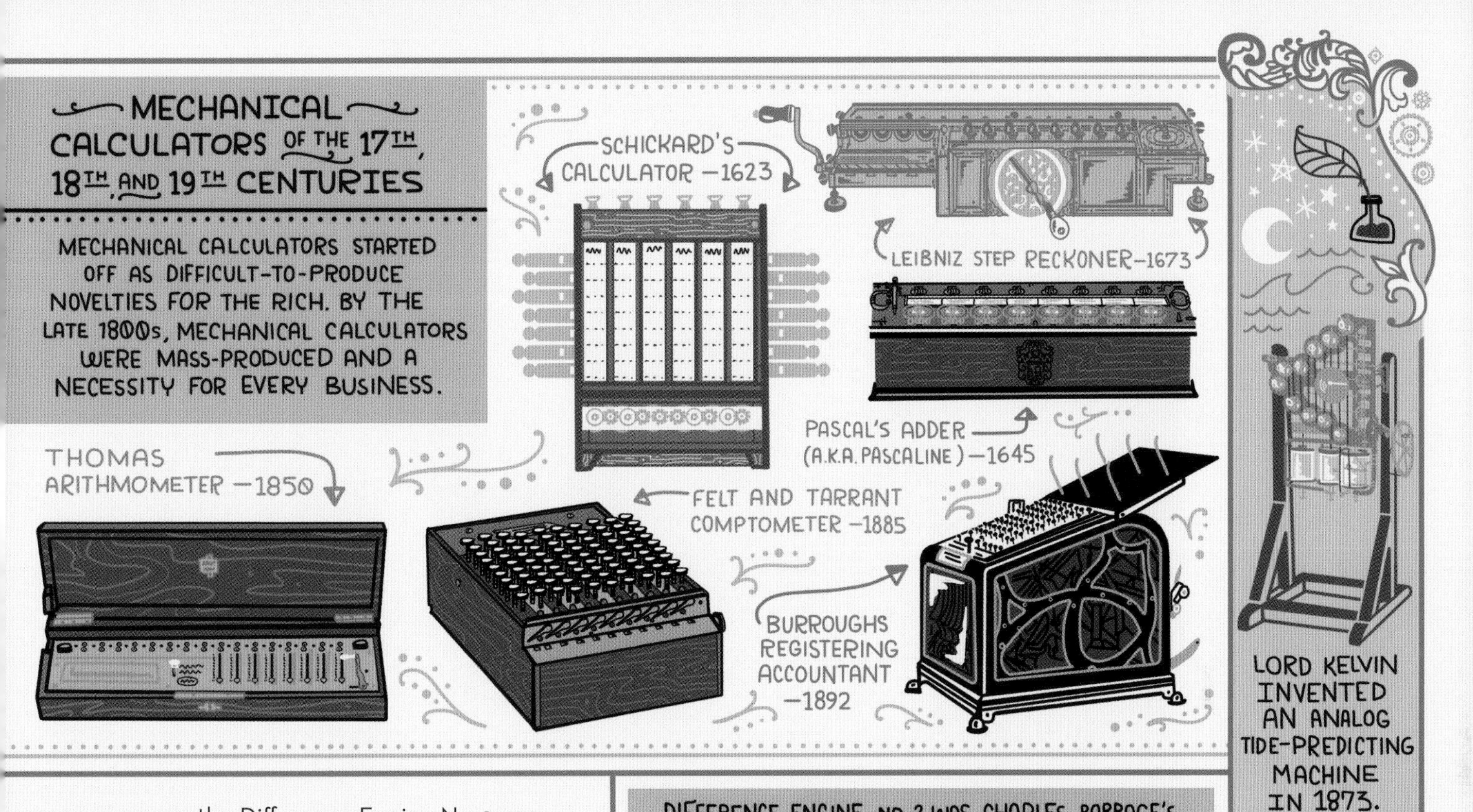

many reasons, the Difference Engine No. 1 was never completed. It was technically complicated to build, with more than twenty-five thousand parts and weighing four tons, and there were rumors that Babbage had fought with his machinist. In 1834, he had spent all his funding for the Difference Engine but had little to show. Around the same time the project stalled, he became distracted with an even better idea that he called the Analytical Engine.

The Analytical Engine was not merely a calculator; it could be programmed to solve any kind of mathematical problem. Inspired by the punch cards used to program mechanical looms in factories (see page 36), Babbage envisioned that punch cards could be used to program the machine and to store information. This general-purpose machine had many of the same functions as a modern computer. Babbage continued his work, funding the Analytical Engine himself and going on to design Difference Engine No. 2 from 1847 to 1849. Although none of Babbage's machines was completed during his lifetime, the Analytical Engine is recognized as the first real dream of a programmable "thinking" machine, and Babbage's work inspired an entire generation of computer scientists.

DIFFERENCE ENGINE NO. 2 WAS CHARLES BABBAGE'S STREAMLINED DESIGN. IT WAS FINALLY AND COMPLETELY BUILT IN 2002 BY THE SCIENCE MUSEUM IN LONDON.

IT TOOK 17 YEARS TO BUILD WITH 8,000 PARTS, WEIGHS 5 TONS, AND IS 11 FEET LONG.

IN 1865, THE PANTELEGRAPH MACHINE WAS USED TO SEND IMAGES, MUCH LIKE A FAX MACHINE.

THE "QWERTY" ARRANGEMENT OF TODAY'S KEYBOARDS WAS FIRST USED IN 1874 ON A TYPEWRITER BY REMINGTON.

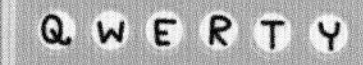

THE SOCIAL SECURITY ACT OF 1935 MEANT MILLIONS OF PUNCH CARDS HAD TO BE ORGANIZED.

IBM CREATED THE TYPE 77 COLLATOR, WHICH PROPELLED BUSINESS TOWARD THE END OF THE GREAT DEPRESSION.

THE UNITED STATES CENSUS

Compared with Europe, the United States was late to the Industrial Revolution. But by the 1880s, both manufacturing and the population of the United States were growing fast—too fast for the U.S. Census to be processed by hand. The census not only counts the number of people living in the country but also gathers such data points as marital status, occupation, age, gender, and race. The U.S. Constitution requires that a census be completed every ten years. When it was time to tabulate the results of the 1880 census, the population had become so large that it took nearly eight years to process the data using the available rudimentary tallying machines and hand counting. Meanwhile, the population continued to grow. It became clear that the 1890 census would be impossible to complete before 1900. The government needed a new way to store and sort this massive amount of data!

In 1888, the U.S. government held a competition to see who could build the fastest machine to process data and, in doing so, earn the lucrative 1890 census contract. Statistician Herman Hollerith won the contract with his electromechanical tabulating machine. He was inspired by punch cards after he noticed that train conductors punched holes in tickets, marking off a passenger's eye color and other descriptors (so the ticket couldn't be reused by a different person). Punch cards were key to storing and sorting data automatically. While other machines took nearly two days to prepare data for tabulation, Hollerith's took only five and a half hours! Using Hollerith's tabulator, trained clerks processed more than sixty million punch cards in two and a half years, saving the U.S. government millions of dollars. This success helped spur an entire industry of punch-card machines that automated data collecting.

WOMEN IN THE WORKFORCE

THE RISE OF NEW TECHNOLOGY MEANT THAT MORE WOMEN ENTERED THE WORKFORCE AS TELEGRAM AND TELEPHONE OPERATORS AND AS HUMAN COMPUTERS.

THE CLERKS HIRED TO OPERATE THE TABULATING MACHINE FOR THE 1890 U.S. CENSUS WERE MOSTLY WOMEN.

TELEPHONE OPERATORS

CENSUS CLERKS

WOMEN HAVE BEEN AN INTEGRAL PART OF MODERN COMPUTER HISTORY SINCE THE BEGINNING!

THE BUSINESS MACHINE

American businesses used electric tabulators during the 1910s and '20s. Calculating and recording payroll, inventory, invoices, and employee attendance all became easier with punch cards and specific machinery. In 1896, Herman Hollerith founded the Tabulating Machine Company, which specialized in punch-card sorting technology. After a series of mergers it was renamed the International Business Machines (IBM) Corporation in 1924.

IBM would rent out the machinery that businesses needed—such as commercial scales, industrial time recorders, and tabulators—while selling single-use punch cards made specifically for those same machines. This is often referred to as the "razor and razor blades model," and it allowed IBM to survive the Great Depression while many other companies went out of business. Over the century, as technology and computing advanced, IBM would continue to be a powerful player in the story of computer history.

IMPACT OF THE ERA

The Industrial Revolution radically shifted the global economy from agricultural labor to manufacturing, which created the need for work that resembled the modern desk job. Governments funded large-scale human computer projects and encouraged inventions such as the electronic tabulator. The automation of data collection became a powerful tool for managing exploding populations and business ambitions. The future favored nations that could pass off this mathematical grunt work to electrical machines. By the end of this era, a few people with deep insight could see the grand potential these machines contained. The groundwork for modern computing was laid.

IMPORTANT INVENTIONS

THE JACQUARD LOOM—1804

The Jacquard loom transformed the way textiles were made and influenced the development of computers. Weaving complex patterns by hand takes a lot of time and requires repetitive labor. Merchant and weaver Joseph Marie Jacquard transformed the textile industry when, in 1804, he invented a mechanical attachment for looms that could automatically weave any pattern. The pattern was "programmed" by a long piece of stiff card stock with holes punched through it. This was the precursor to the punch cards that would be used to create future complicated computer programs.

THE FIRST COMPUTER PROGRAM AND THE ANALYTICAL ENGINE—1843

Mathematician and poet Ada Lovelace became familiar with Charles Babbage's work at the age of seventeen, when Babbage showed off a small working piece of his Difference Engine at a party. The two then began a long working friendship and collaboration. Lovelace was especially inspired by the designs for the Analytical Engine, which essentially was a general-purpose computer. In 1843, she published a translation of a French article on the Analytical Engine and included her own commentary and notes. Lovelace wrote about the potential for this machine to do more than just math calculations, recognizing that it had the capacity to do anything a programmer could imagine (even create music!). As part of her notes, Lovelace also wrote an algorithm that the Analytical Engine could run. Historians now consider this to be the first-ever computer program. Her notes became one of the most important documents in computer history.

CARD READER

CLERKS CLOSED THE PLATES AND, WHEREVER THERE WAS A HOLE IN A CARD, A PIN WOULD DROP INTO A SMALL WELL OF MERCURY, COMPLETING THE CIRCUIT AND ADDING TO THE COUNTER ON A SPECIFIC DIAL.

SORTING TABLE

AFTER THE MACHINE REGISTERED A CARD, A SPECIFIC DRAWER OPENED, SHOWING THE CLERK WHERE THE PUNCH CARD SHOULD BE PLACED.

HOLLERITH TABULATOR DIALS

EACH OF THE 40 DIALS REPRESENTED A DIFFERENT DATA ITEM.

PILES OF PUNCH CARDS

A PANTOGRAPH

U.S. CENSUS CLERKS PUNCHED THE DATA FOR EACH CITIZEN INTO A CARD.

HOLLERITH'S TABULATOR – 1888

The first electromechanical tabulating machine was used for the 1890 census (installed in 1888). Dials counted the number of holes in a specific spot on the punch card, and the specific position of the holes on the cards was also used to sort the cards automatically. This made it super-easy to generate statistics. For example, a clerk could quickly find out how many firemen were married, owned their own homes, and were over the age of twenty-five. Each time the machine read a card, a bell would ring, letting the operator know to record the data. An experienced clerk could process about eighty cards in a minute, making the process ten times faster than sorting by hand.

INFLUENTIAL PEOPLE

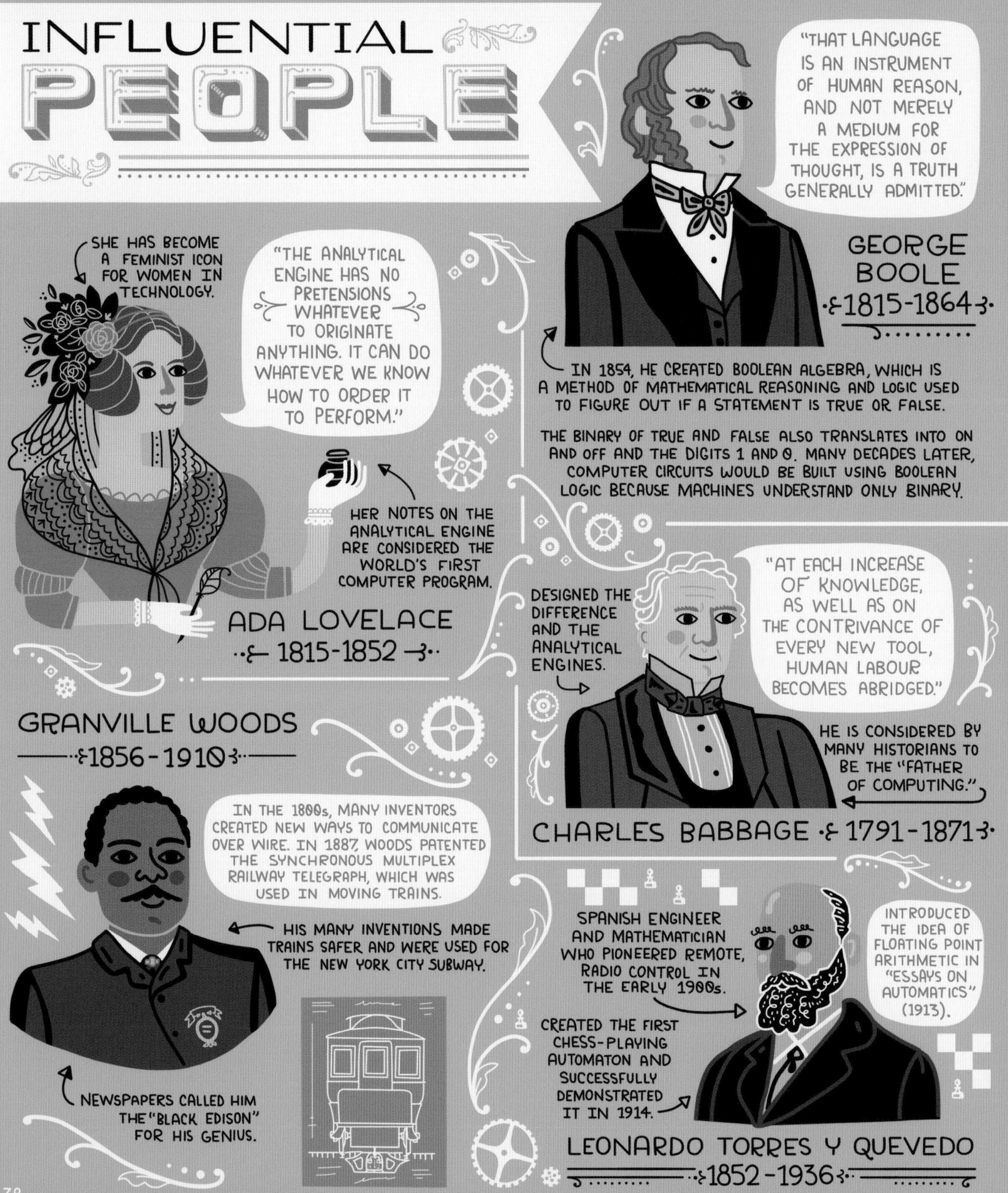

FEATURED BIOGRAPHIES

HERMAN HOLLERITH 1860-1929

Herman Hollerith won the contract for the 1890 census with his futuristic "electromechanical punched card tabulator," which was patented in 1889. Hollerith's tabulator was a huge success that would forever change how information was processed. Hollerith started the Tabulating Machine Company in 1896, renting his machines to governments around the world. Because he had a monopoly on tabulating machines, he became greedy and raised his rental prices. This increase made rented machines too expensive for the U.S. Census Bureau. For the 1910 census, the government developed its own tabulating machines and almost (but not quite) broke patent law to do it.

In 1912, Hollerith sold his company, which then became CTR, but he remained its chief consultant. He was stubborn when it came to improving upon his original designs, and CTR soon faced financial trouble and needed new innovations. In 1914, CTR hired Thomas J. Watson Sr., who began making improvements to the company by creating research teams and implementing a new sales strategy. Hollerith stayed with the company until he retired to farm life in 1921, stating that he was "fully occupied with boats, bulls, and butter."

Hollerith saw himself as a statistical engineer, but his tabulating machines ended up being used in ways far beyond how he could ever have imagined. His inventions were the foundation of how data would be processed for the next hundred years.

THOMAS J. WATSON SR. 1874-1956

Thomas J. Watson Sr. started his career as a traveling salesman, selling pianos and organs from a horse-drawn wagon. In 1895, he began working as a salesman for NCR (National Cash Register) Corporation. NCR's president, John Henry Patterson, was an eccentric boss. He often forced employees to go horseback riding and gave bonuses or fired employees depending on his mood. But he created a strong corporate culture by giving speeches, coining slogans, and offering perks to incentivize salesmen. Later, this would inspire Watson's approach to leadership at IBM. Watson ran NCR's sales school and coined the slogan THINK. Then, after Watson became the general manager of NCR in 1911, Patterson fired him in one of his fits. When Watson left, he took his THINK mantra with him and got a job at CTR (which would become IBM under his guidance). Within eleven months, Watson became the company president and created a team devoted solely to research and invention, giving IBM a competitive edge.

Under Watson's leadership, IBM held a virtual monopoly on the business-machine market. During the 1930s, he aggressively and thoughtlessly pursued international trade, including supplying census machines to Nazi Germany. Those same machines would end up being used in the Holocaust. Like many titans of industry in the 1930s, Watson foolishly believed in the liberalizing force of commerce in the face of state-sponsored hatred.

Although Watson did not create any new technology himself, through his (sometimes questionable) business tactics, he built one of the most powerful technology companies in the world. During World War II, under Watson's leadership, IBM helped build one of the first computers for the U.S. military, the Mark I. He retired from IBM a month before his death, handing off the reins of his corporate empire to his son, Thomas J. Watson Jr.

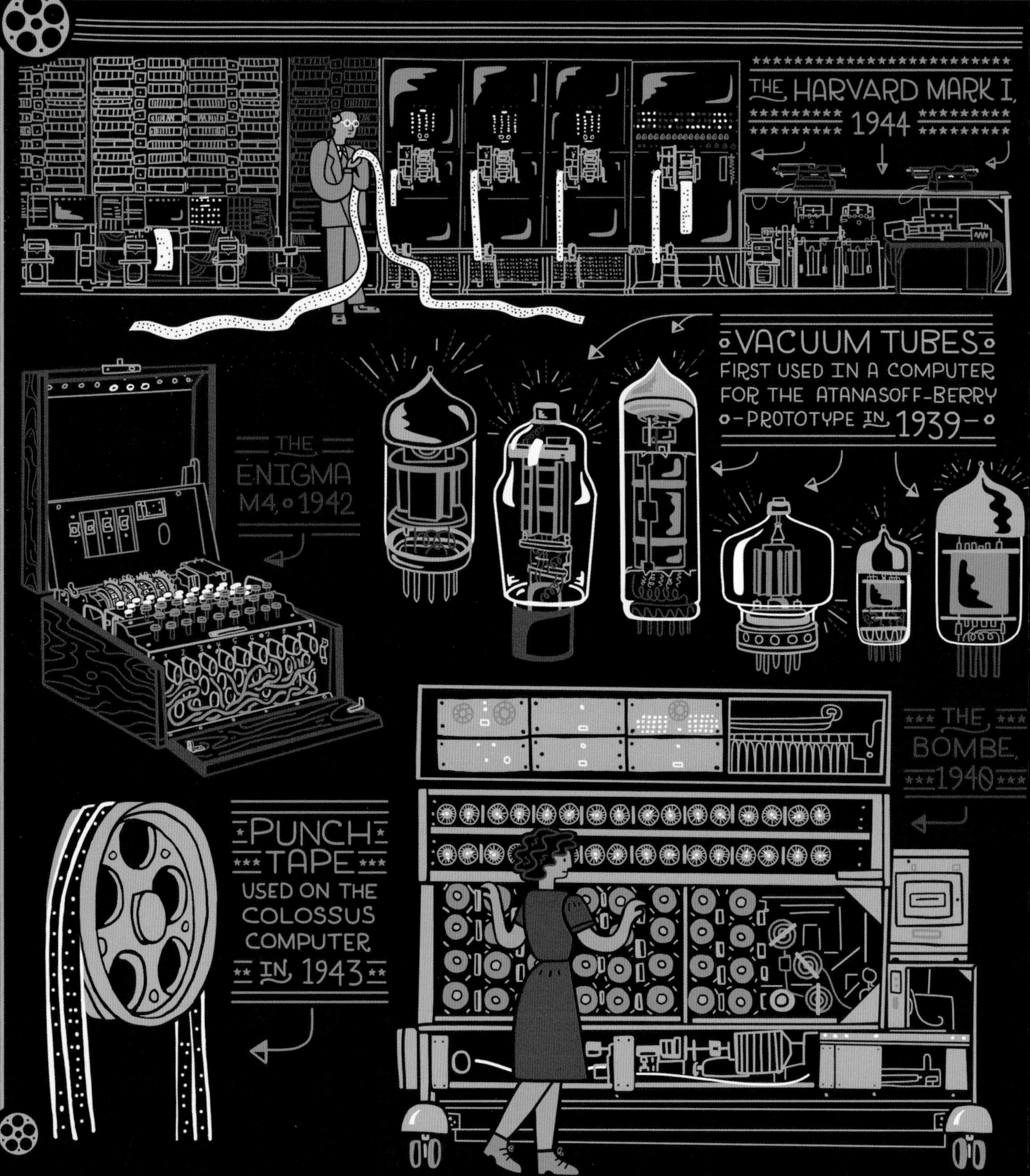
THE HARVARD MARK I, 1944
VACUUM TUBES
FIRST USED IN A COMPUTER FOR THE ATANASOFF-BERRY PROTOTYPE IN 1939
THE ENIGMA M4, 1942
THE BOMBE, 1940
PUNCH TAPE
USED ON THE COLOSSUS COMPUTER IN 1943

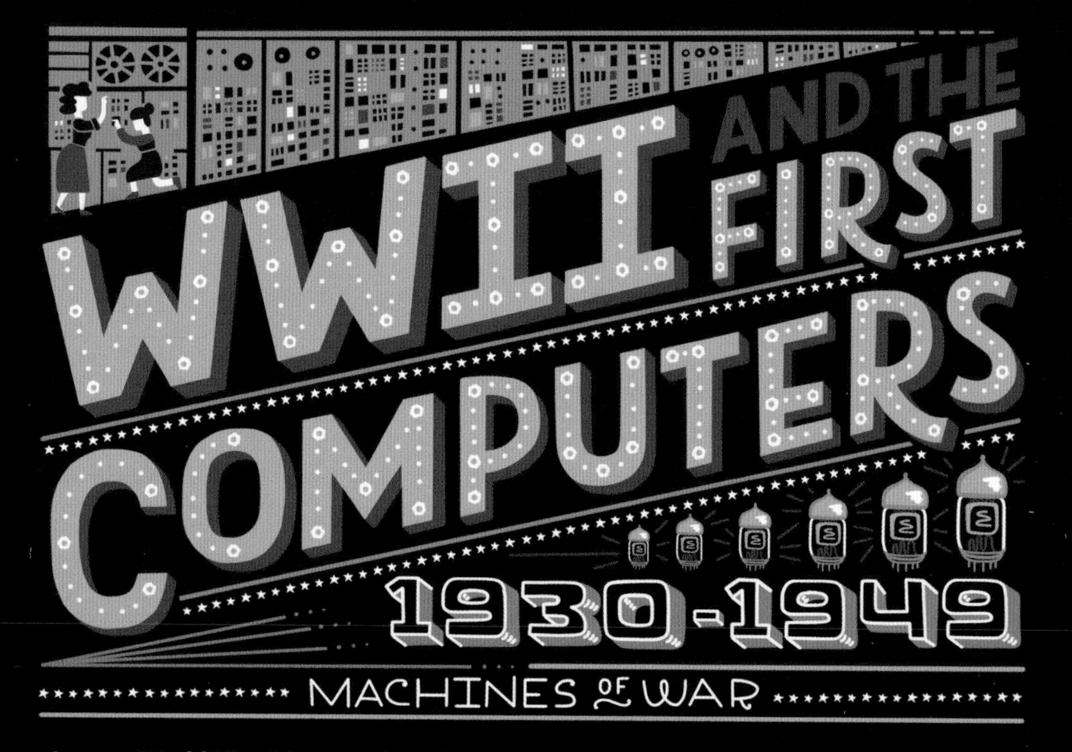

In 1939, World War II began when Nazi Germany invaded Poland. Germany wanted to dominate the entire world and began invading Europe, while undertaking a state-sponsored genocide of Jewish, Roma, disabled, and LGBTQ+ populations; this would become known as the Holocaust. The ensuing war was a fight against Fascism that lasted until 1945, dividing the globe into the Axis powers (Germany, Japan, and Italy) and the Allied powers (the United Kingdom, United States, Soviet Union, and China). World War II was fought at a massive scale, mobilizing an estimated seventy million troops worldwide, and required technology such as ballistic trajectories, radar systems, and code breaking that necessitated the need for computation on a grand scale. It was this war effort—which established large-scale, military-funded technology projects—that would create the first generation of programmable computers.

Massive amounts of money and human brain power were spent on top-secret government projects to create "mechanical minds" that would help break secret codes and build bombs. Colossus in the United Kingdom and the Harvard Mark I in the United States were some of the very first computers. These gigantic, dimly lit, hot-to-the-touch machines filled large rooms with whizzing punch tape, clinking parts, and blinking lights. These war machines helped with calculations that were too complicated and time-consuming for people to do by hand. Computers helped the Allies win World War II, and this cemented computer technology in the arsenal of warfare.

1936 MODEL K ADDER

Bell Labs scientist George Stibitz created a simple logic circuit that could add two binary digits. He built it on his kitchen counter using scrap relays and metal from tin cans and showed that Boolean logic could be used to design computers.

THE HP GARAGE, PALO ALTO, CA

DAVID PACKARD

BILL HEWLETT

1938 THE START OF SILICON VALLEY

The "Birthplace of Silicon Valley" is considered to be the HP (Hewlett-Packard) Garage, in Palo Alto, California, where Bill Hewlett and David Packard built radios to start their company. In the 1970s, the southern San Francisco Bay Area would be nicknamed "Silicon Valley" because of its many computer companies!

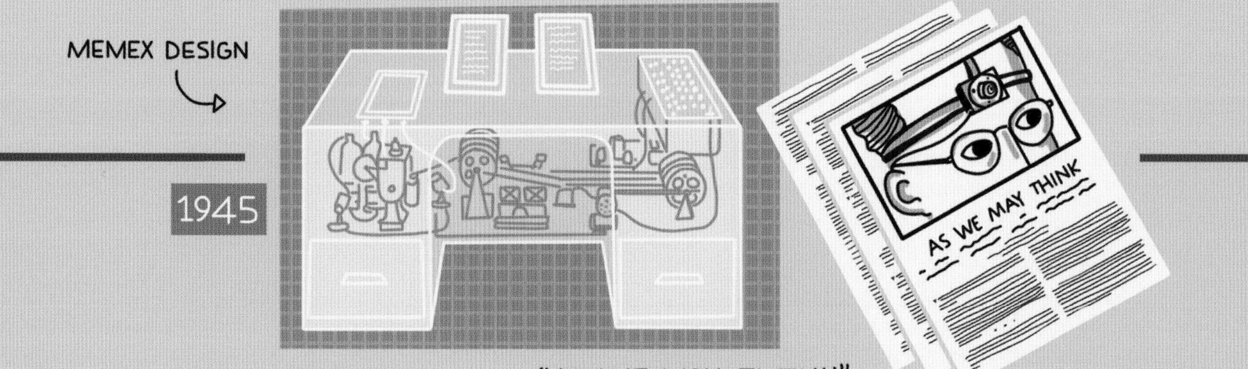

1945 "AS WE MAY THINK"

American engineer and science administrator Vannevar Bush published his influential speculative essay about a memory expander, called the memex (a combination of "memory" and "index"). In it, he described such future technology as an online encyclopedia, hypertext, and essentially the internet itself—decades before they would be invented.

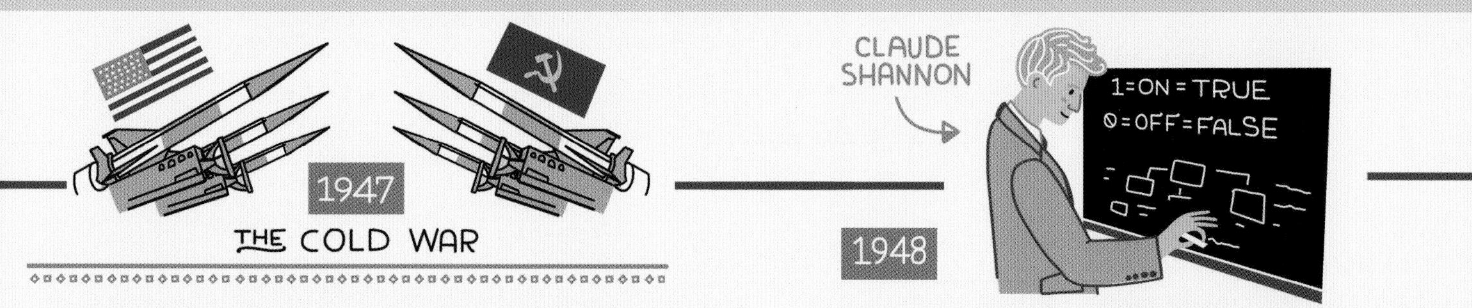

1947 THE COLD WAR

After World War II ended, a period of geopolitical tension known as the Cold War began between the United States and the Soviet Union. Author George Orwell described a *cold war* as "two or three monstrous superstates, each possessed of a weapon by which millions of people can be wiped out in a few seconds."

1948 THE BIT

In Claude Shannon's paper "A Mathematical Theory of Communication," he defines the *bit*. A bit is represented by a single binary digit, a 0 or a 1. It is the smallest and most fundamental unit of information.

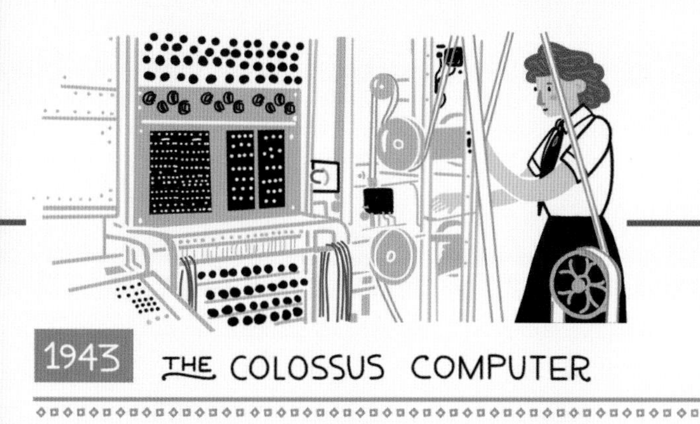

1943 THE COLOSSUS COMPUTER

The U.K. military constructed the Colossus computer from 1943 to 1945 at Bletchley Park, an ultra-secret military campus.

THE HARVARD MARK I

The United States built one of the first programmable computers, the Harvard Mark I. It was operational for the U.S. war effort and began calculations for the Manhattan Project.

1946 ENIAC GOES PUBLIC

ENIAC, the first programmable, general-purpose electronic computer, was revealed to the public. The press was astounded by its "mechanical mind."

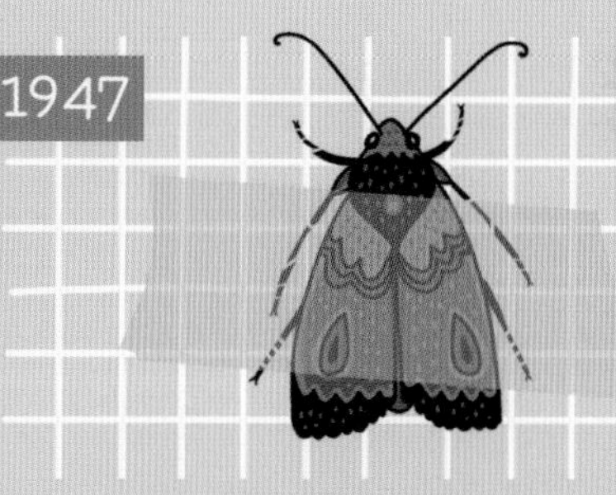

FIRST "COMPUTER BUG"

The Harvard Mark I and II were physically hot and attracted insects. In 1947, a moth that got inside the Mark II caused a hardware malfunction. It became known as the first "computer bug."

WAS BASED ON A PARTY GAME WHERE PLAYERS HAD TO GUESS THE GENDER OF THE OTHER PERSON, WHOM THEY COULDN'T SEE.

THE TURING TEST

British mathematician and cryptologist Alan Turing developed a method to tell whether a computer was truly "intelligent." An interrogator asked both a computer and a person similar questions and then guessed which of the two was human. If the computer tricked the interrogator, then it was considered intelligent. This was a crucial theory in the development of AI, and became known as the "Turing Test."

STORIES FROM HISTORY

World War II was fought on multiple fronts, including ultra-secret laboratories. It was a race to see which side would develop the technology first—the fastest-firing guns, the most advanced radar system, the most complex encryption and code-breaking methods, and the biggest bombs. Winning this war required computation on a scale that was too great for human computers and mechanical calculators alone, so with great haste and immense secrecy, the United Kingdom and the United States each developed wartime computers to gain the competitive edge.

SCI-FI AUTHOR ISAAC ASIMOV WROTE "RUNAROUND" (1942) AND NOVELIZED SHORT-STORY COLLECTION *I, ROBOT* (1950), WHICH INSPIRED FUTURE AI RESEARCHERS.

BLETCHLEY PARK

Throughout the war, the United Kingdom was under attack from Nazi Germany. Information is the lifeblood of any army, and the British military knew they needed to crack Nazi secret messages. As bombs rained down on British cities, the government assembled a clandestine code-breaking team in the countryside at Bletchley Park.

The German military communicated through coded messages, created by thousands of Enigma machines. The messages were easy to intercept but impossible to understand without the cipher key, which the Nazis would change every single day. There were more than 150 quadrillion combinations and only twenty-four hours to crack each code. Within the campus at Bletchley Park, mathematical genius Alan Turing led a special code-breaking team.

Years earlier, in 1938, the Polish Cipher Bureau built a machine called the Bomba (named perhaps for its ticking sound) that could decode Enigma machine messages. During World War II, new Enigma machines made the Bomba obsolete. Based on the older Polish machine, Turing developed a much-improved code breaker called the Bombe. The Bletchley Park team exploited a flaw in the Enigma machine—it never used the same letter twice to represent itself in a coded message. The first two Bombes were named "Victory" and "Agnes," and the Bletchley team built many more for the war effort. But this was still not enough to crack the Lorenz ciphers used by Nazi High Command.

Lorenz ciphers were an even more complicated encryption that utilized twelve different encryption wheels. At Bletchley Park, physicist Tommy Flowers spent eleven months developing a programmable machine with the sole purpose of cracking these Nazi High Command messages. It was called the Colossus and became operational in 1943 as one of the very first electronic computers. Colossus's input was a continuous roll of punched paper tape that moved at about twenty-seven miles per hour and could crack Lorenz ciphers in mere hours instead of weeks. Ten Colossus computers were built for the war effort between 1943 and 1945.

The information retrieved from the Colossi and the Bombes aided in many military operations, including the invasion of Normandy. After the war, the Colossus computers were scrapped and recycled. All the work done at Bletchley Park remained at the highest level of secrecy for decades.

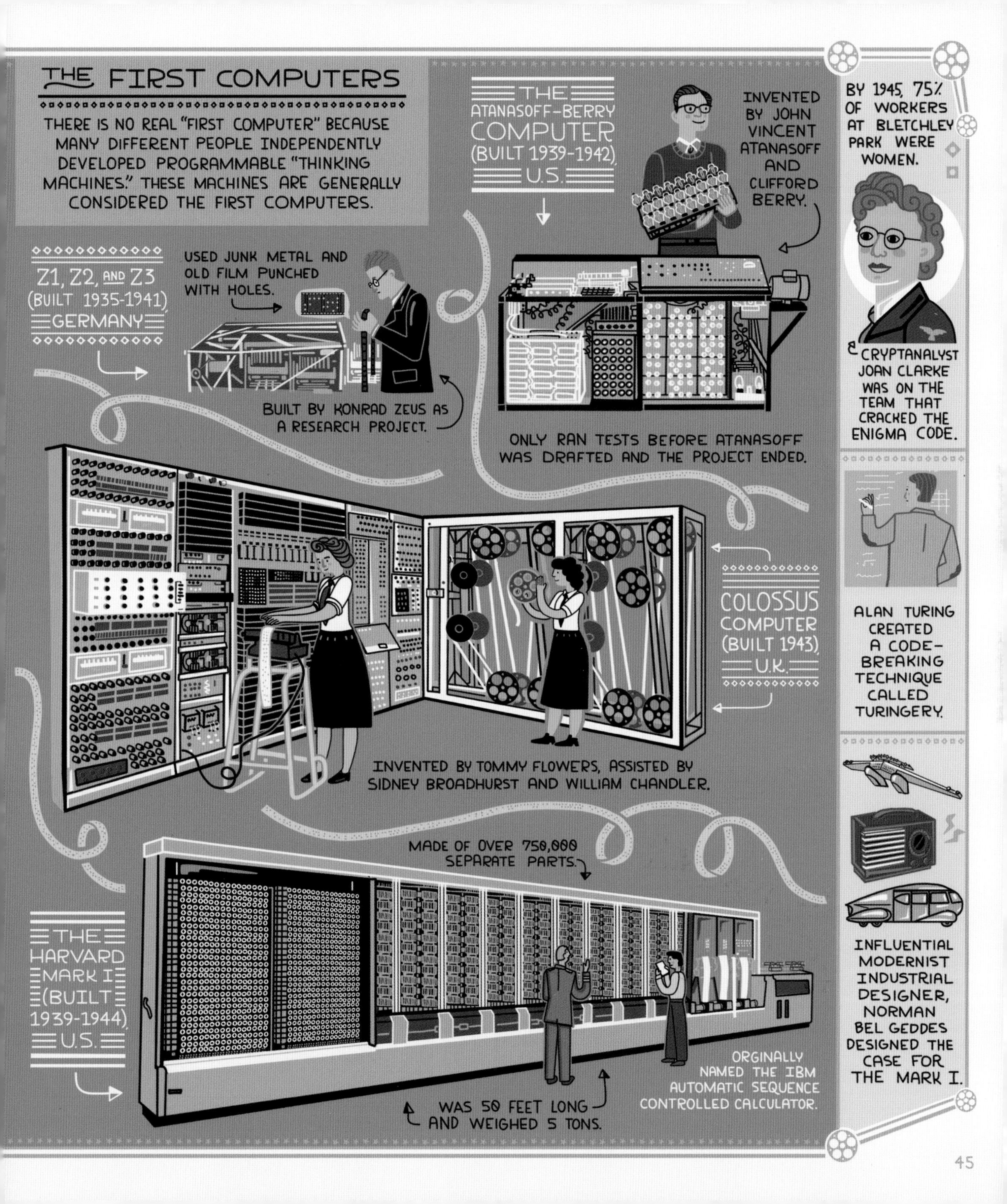
THE FIRST COMPUTERS
THERE IS NO REAL "FIRST COMPUTER" BECAUSE MANY DIFFERENT PEOPLE INDEPENDENTLY DEVELOPED PROGRAMMABLE "THINKING MACHINES." THESE MACHINES ARE GENERALLY CONSIDERED THE FIRST COMPUTERS.
Z1, Z2, AND Z3 (BUILT 1935-1941) GERMANY
USED JUNK METAL AND OLD FILM PUNCHED WITH HOLES.
BUILT BY KONRAD ZEUS AS A RESEARCH PROJECT.
THE ATANASOFF-BERRY COMPUTER (BUILT 1939-1942), U.S.
INVENTED BY JOHN VINCENT ATANASOFF AND CLIFFORD BERRY.
ONLY RAN TESTS BEFORE ATANASOFF WAS DRAFTED AND THE PROJECT ENDED.
COLOSSUS COMPUTER (BUILT 1943), U.K.
INVENTED BY TOMMY FLOWERS, ASSISTED BY SIDNEY BROADHURST AND WILLIAM CHANDLER.
THE HARVARD MARK I (BUILT 1939-1944), U.S.
MADE OF OVER 750,000 SEPARATE PARTS.
WAS 50 FEET LONG AND WEIGHED 5 TONS.
ORGINALLY NAMED THE IBM AUTOMATIC SEQUENCE CONTROLLED CALCULATOR.
BY 1945, 75% OF WORKERS AT BLETCHLEY PARK WERE WOMEN.
& CRYPTANALYST JOAN CLARKE WAS ON THE TEAM THAT CRACKED THE ENIGMA CODE.
ALAN TURING CREATED A CODE-BREAKING TECHNIQUE CALLED TURINGERY.
INFLUENTIAL MODERNIST INDUSTRIAL DESIGNER, NORMAN BEL GEDDES DESIGNED THE CASE FOR THE MARK I.

CODE BREAKING WAS IMPORTANT TO THE U.S. WAR EFFORT.

WILLIAM COFFEE LED A UNIT OF 100 BLACK AMERICAN CRYPTOLOGISTS. THEY DID IMPORTANT WORK WHILE SEGREGATED FROM THE OTHER CODE BREAKERS.

CLICK CLICK CLICK

WHEN THE HARVARD MARK I WAS TURNED ON, IT SOUNDED LIKE "A ROOMFUL OF WOMEN KNITTING."

FIRE CONTROL

The U.S. military needed to calculate the trajectory of long-range guns; this was called Fire Control. Artillery operators couldn't simply aim straight at a target. Instead, they needed to take into account the curvature of the earth, weather, humidity, and wind speed. All of these variables went into differential equations. As far back as World War I, mechanical calculators, called rangekeepers, were used to calculate and direct gunfire on the battlefield and open seas.

American electrical engineer Vannevar Bush and his graduate student Harold Locke Hazen built the Differential Analyzer in 1931. This machine calculated differential equations that were impossible to solve by hand. With impressive speed, it crunched the numbers to simulate earthquakes, understand weather patterns, build electrical networks, and, of course, calculate ballistic trajectories. In 1942, Bush built the improved, fully electric RDA (Rockefeller Differential Analyzer). The RDA was not a computer but was considered one of the most important math machines employed during World War II. It helped calculate firing tables, radar antenna, and equations for the atomic bomb. Bush's RDA was still limited in the kind of equations it could do. The U.S. military needed to build a programmable computer.

THE FIRST MACHINE TO BE CALLED A "COMPUTER" WAS THE MARK I

"BY 'COMPUTER' WE SHALL MEAN A MACHINE CAPABLE OF CARRYING OUT AUTOMATICALLY A SUCCESSION OF OPERATIONS OF THIS KIND AND OF STORING THE NECESSARY INTERMEDIATE RESULTS..."
—GEORGE STIBITZ

THE UNITED STATES BUILDS ITS FIRST COMPUTERS

Howard Aiken, a Harvard graduate student, would pick up where Charles Babbage had left off a century earlier in developing a programmable computer. In 1936, Aiken designed his own large-scale digital calculator. In Harvard's library, Aiken came across Babbage's writings from the nineteenth century, saying that he "felt that Babbage was addressing me personally from the past." Three years later, Aiken began working with IBM, which supplied funding and its best engineers to build his programmable computer. When Aiken joined the U.S. Navy in 1941, his machine became a special military project. It was renamed the Harvard Mark I and was completed in 1944.

First-generation computers such as the Mark I were slow and "dumb" by modern standards, but much quicker than doing calculations by hand. A computer's speed depends on how quickly it can turn the flow of electricity within its circuitry on and off. The Mark I's mechanical switches had to physically move; this took time and meant that its switches could wear down and break. Once an operation was set, it took the Mark I many hours—sometimes days—to complete a calculation, which Aiken said was the machine "makin' numbers." The Mark I was used until 1959 for many military calculations, including radar development, surveillance camera lenses, and torpedo design. Aiken continued to improve the Harvard Mark series for the U.S. military, leading the design teams for the Mark I through IV (the last one was built in 1952).

While the Mark I was being built, another top-secret computer project was underway, deep in the basement of the Moore School of Electrical Engineering at the University of Pennsylvania. From 1943 to 1945, physicist John Mauchly and inventor J. Presper Eckert led a team in developing the first large-scale computer to run at electronic speed. It would be called ENIAC (Electronic Numerical Integrator and Computer). Unlike the Mark I's mechanical switches, ENIAC used vacuum tubes to switch between on and off states; this meant it had no moving parts and was much faster. ENIAC was completed in 1945, months after World War II ended. Scientists announced it could calculate trajectory faster than a speeding bullet. (Learn more about ENIAC on page 49.) When the military decided to show ENIAC to the public, a headline read "Blinkin' ENIAC's a Blinkin' whiz!" and a newsreel said "It's the world's first electronic computer. Right now, it's solving mathematical problems for the U.S. Army. But who knows? Someday a machine like this may check up on your income tax." The U.S. military used ENIAC for the next ten years, and it was the first of many military machines used during the Cold War.

COMPUTERS AND THE BOMB

Both the Harvard Mark series and ENIAC were used on the Manhattan Project (active from 1942 to 1946). There was fear that the German government was developing its own nuclear weapons, and the Manhattan Project was a highly classified, national priority to create an atomic bomb. In 1945, the U.S. military dropped atomic bombs on the Japanese cities of Hiroshima and Nagasaki, killing more than two hundred thousand people, most of whom were civilians. Many people mark this brutal display of power as the end of World War II, although historians still debate whether dropping those bombs was necessary to end the war.

IMPACT OF THE ERA

The first-generation computers were far from the intimate devices that we carry in our pockets today. They filled entire rooms, took a ton of electricity to run, and were physically taxing to program. These room-size machines blinked and ticked while they calculated casualties and aimed missiles. They were used on secret projects and were operated only by trained scientists and military personal. Even after the war, computers' size and price would make them out of reach of the general public for decades.

The United States became a leader in technology after World War II for many reasons. Most of Europe and many other parts of the world had been devastated from bombings and warfare, while the continental United States remained intact. Unlike the United Kingdom, the United States did not keep their military technology a complete secret, and many U.S. projects were funded by universities and private companies. America's postwar boom created a new middle class and many corporate desk jobs, which would later use technology that was conceived during the war.

AMERICAN WOMEN HELPED THE WAR EFFORT IN FACTORIES, SHIPYARDS, AND "HUMAN COMPUTER" PROJECTS FOR WARTIME CALCULATIONS.

DURING WWII, ACTRESS HEDY LAMARR AND COMPOSER GEORGE ANTHEIL INVENTED A VERSION OF FHSS (FREQUENCY-HOPPING SPREAD SPECTRUM) COMMUNICATION.

INITALLY MADE TO STOP TORPEDO HIJACKINGS, THEIR WORK LATER BECAME THE BASIS FOR BLUETOOTH AND WI-FI TECH.

IMPORTANT INVENTIONS

THE TRANSISTOR – 1947

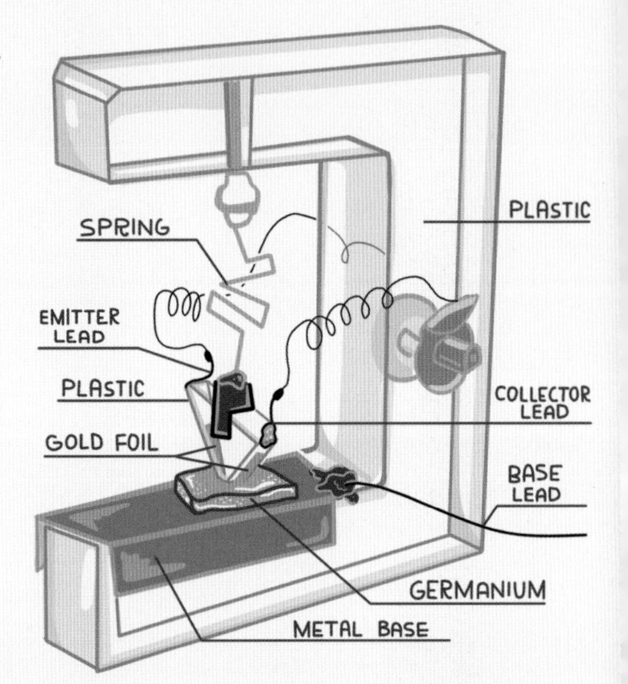

The basis of a computer's circuity is the ability to switch between ON and OFF states. Computers, like the mechanical Harvard Mark I, had moving parts that were unreliable and slow. The ENIAC and Colossus computers used vacuum tubes that were fragile and burned out quickly. There had to be a better way!

In 1947, John Bardeen and Walter Brattain at Bell Labs were trying to create a device to amplify ultra-high-frequency radio waves. They began experimenting with a semiconducting crystal called germanium. They discovered that by attaching two gold contacts (placed very closely but not touching) and then applying electricity to the germanium, they could amplify a signal. This became known as a point-contact transistor.

Transistors use a fraction of the electricity of vacuum tubes and are smaller and much more durable. The transistor would replace vacuum tubes used in radios and TVs. The point-contact transistor was also able to switch an electric signal on and off, making it a solid-state electronic switch. In 1953, at the University of Manchester, the first fully transistorized computer was built.

RADAR – 1934

It had been known since the 1880s that solid objects reflect and "bounce" radio waves, similar to how your voice echoes in a cave! With this principle, in 1934, at the U.S. Navy Research Lab, scientists demonstrated a working radar (RAdio Detection And Ranging) using pulses of radio waves to track a plane flying a mile away over the Potomac River. A faint green dot on an oscilloscope showed its location.

The United States and the United Kingdom shared radar technology during World War II. In the Battle of Britain, the Royal Air Force (RAF) was outnumbered by German aircraft, but thanks to Britain's superior radar system, the smaller RAF was able to see the enemy coming from 100 miles away and intercept them.

Radar and computer history are tightly wrapped together. Many of the first computer displays were repurposed radar scopes. Technology such as transistors and delay line memory were first developed for radar and then used in early computers. Today, technology based on radar is found in billions of devices. It is crucial for air traffic control, weather monitoring, space exploration, and in smartphones!

DELAY LINE MEMORY IN EARLY COMPUTERS USED TECHNOLOGY ORIGNALLY MADE FOR RADAR.

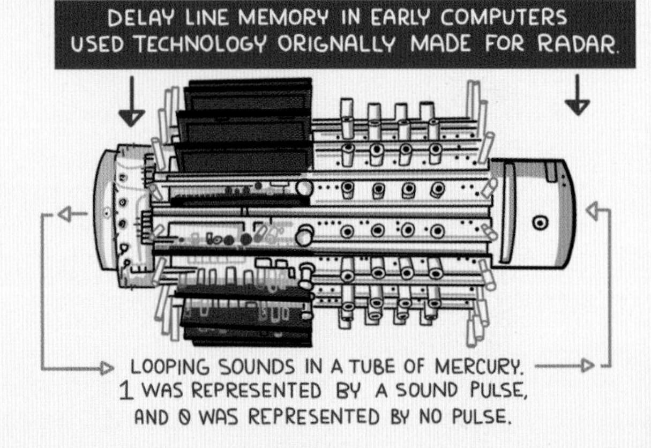

LOOPING SOUNDS IN A TUBE OF MERCURY. 1 WAS REPRESENTED BY A SOUND PULSE, AND 0 WAS REPRESENTED BY NO PULSE.

ENIAC WAS 30 TONS, 8 FEET TALL, 3 FEET DEEP & 100 FEET LONG.

AT LEAST 1 OF ITS 18,000 VACUUM TUBES HAD TO BE REPLACED ABOUT EVERY OTHER DAY.

J. PRESPER ECKERT

JOHN MAUCHLY

SETTING SWITCHES ON ONE OF ENIAC'S FUNCTION TABLES.

ENIAC OPERATIONAL — 1945–1955

ENIAC was the first large-scale general-purpose computer to run at electronic speed. It read IBM punch cards for input and output and was programmed with three function tables that looked like telephone switchboards and each had 1,200 ten-position rotary switches. The machine was imposing, filling a 100-foot room, and programming it was a physically laborious task. Its 18,000 vacuum-tube lights blinked as it computed. In its ten years of operation, ENIAC did more calculations than had been done in all of human history combined to that point.

INFLUENTIAL PEOPLE
TOMMY FLOWERS
1905-1998
WORKED AS A CODE BREAKER AT BLETCHLEY PARK AND WITH ALAN TURING ON CRACKING THE ENIGMA MACHINE. IN 1944, FLOWERS LED THE TEAM THAT BUILT THE COLOSSUS COMPUTER.
AFTER THE WAR, HE WAS FORCED TO DESTROY ALL EVIDENCE OF COLOSSUS AND BURN HIS DESIGNS.
COLOSSUS AIDED IN THE PLANNING OF D-DAY.
"I VISUALIZE A TIME WHEN WE WILL BE TO ROBOTS WHAT DOGS ARE TO HUMANS, AND I'M ROOTING FOR THE MACHINES."
CLAUDE SHANNON
1916-2001
HE FIGURED OUT HOW BOOLEAN LOGIC COULD BE USED FOR CIRCUITRY.
HE WAS NICKNAMED "THE FATHER OF INFORMATION THEORY" BECAUSE OF HIS 1948 PAPER "A MATHEMATICAL THEORY OF COMMUNICATION."
SHANNON WAS AN IMPORTANT CRYPTANALYST FOR THE U.S. IN WWII. HIS 1949 PAPER "COMMUNICATION THEORY OF SECRECY SYSTEMS" IS BASED ON THE SECRET WORK HE DID DURING WARTIME.
VANNEVAR BUSH
1890-1974
"AS LONG AS SCIENTISTS ARE FREE TO PURSUE THE TRUTH WHEREVER IT MAY LEAD, THERE WILL BE A FLOW OF NEW SCIENTIFIC KNOWLEDGE TO THOSE WHO CAN APPLY IT TO PRACTICAL PROBLEMS."
IN 1945, BUSH WROTE INFLUENTIAL ESSAY "AS WE MAY THINK."
AS DIRECTOR OF THE OFFICE OF SCIENTIFIC RESEARCH AND DEVELOPMENT, BUSH ORGANIZED MILITARY SPENDING TO FOCUS ON SCIENCE TO HELP WIN WWII. HE OVERSAW AND ORGANIZED THE MANHATTAN PROJECT.
IN 1950, BUSH STARTED THE NSF (NATIONAL SCIENCE FOUNDATION) FOR PEACETIME GOVERNMENT-FUNDED RESEARCH.
"DON'T WORRY ABOUT PEOPLE STEALING YOUR IDEAS. IF YOUR IDEAS ARE ANY GOOD, YOU'LL HAVE TO RAM THEM DOWN PEOPLE'S THROATS."
HE PUBLISHED NUMEROUS PAPERS ON ELECTRONICS AND DATA PROCESSING.
HE LED THE TEAM THAT BUILT THE HARVARD MARK I-IV.
HOWARD AIKEN
1900-1973
HE WAS ONE OF THE LEADERS IN THE WAR EFFORT'S COMPUTER SCIENCE PROGRAM.

FEATURED BIOGRAPHIES

ALAN TURING 1912-1954

English mathematician Alan Turing is considered one of the most important figures in computer history. In 1935, Turing conceptualized the modern computer. He argued that given enough time and memory, this machine could simulate any algorithm, no matter how complicated. A "Turing machine" essentially describes a powerful type of computer. Today, smartphones and laptops are considered "Turing complete."

Turing earned his PhD at Princeton University in 1938 and returned to the United Kingdom to work in cryptanalysis. During World War II, he led top-secret code-breaking projects at Bletchley Park. The intelligence collected by his team was crucial to an earlier ending of the war. Later, Turing formed important ideas about AI in his 1950 paper "Computing Machinery and Intelligence," posing the ultimate question, "Can machines think?" He imagined a future where computers had enough complexity that they could appear as intelligent as a human. Turing devised a test that asked conversational questions to both a computer and a human. If the computer could trick the interrogator into thinking it was human, then it would be considered "intelligent." This is called the Turing Test and is foundational to the development of AI.

Turing was a gay man at a time when that was a crime. In 1952, police spied on Turing's romantic life, and he was faced with the cruel choice between jail or probation with forced hormone treatment. To continue his work, Turing chose the latter, however, the hormones caused him to suffer from depression. It is generally accepted by historians that he took his own life in 1954. In 2013, he posthumously received a royal pardon and is rightfully remembered as a war hero. Having possessed one of the greatest minds of all time, the Turing Award in computer science is named in his honor.

KATHLEEN ANTONELLI (1921-2006)

FRANCES SPENCE (1922-2012)

"WE WERE SURE THAT THIS MACHINE COULD DO ANYTHING WE WANTED IT TO DO. WE WERE VERY COCKY ABOUT THAT, SO WE SET ABOUT TO TRY AND MAKE IT DO IT!"

MARLYN MELTZER (1922-2008)

JEAN BARTIK (1924-2011)

FRANCES "BETTY" HOLBERTON (1917-2001)

RUTH TEITELBAUM (1924-1986)

THE ENIAC WOMEN

After ENIAC was completed in 1945, there was the daunting task of figuring out how to program the machine. This job would be done by six female human computers. At first, the ENIAC women didn't even have the security clearance to see the top-secret computer in person. At the time, no programming tools existed, so they used wiring block diagrams to figure out the logic to program the machine. When they finally were allowed to see ENIAC, they programmed it by physically plugging in hundreds of cables and setting three thousand switches. They created methods to debug the computer and are credited with creating one of the first sorting algorithms and the first software application.

In 1946, ENIAC was made public by the press. Despite their tremendous accomplishment, the ENIAC women were not publicly credited for their work. Jean Bartik recalls, "We were never treated as though we knew anything. When the press came, they had us act like models and pretend to set switches. We were never considered a part of history." These women went on to teach the first generation of computer programmers. Bartik worked on the BINAC and UNIVAC. Frances "Betty" Holberton worked on the project that developed programming language COBOL. Despite being unrecognized in their time, the ENIAC women changed the course of computer history and are now recognized as pioneers!

CORE MEMORY FIRST USED IN COMPUTERS IN 1953
MAGNETIC TAPE FIRST USED TO RECORD COMPUTER DATA IN 1951
PART OF THE IBM SYSTEM/360 INTRODUCED IN 1964
STAR TREK PREMIERED ON TV IN 1966
IBM
360
MOON LANDING, 1969
THE UNIVAC SYSTEM BROCHURE FROM 1955
The UNIVAC SYSTEM

After World War II, the United States and the Soviet Union emerged as the world's two superpowers. Their ideological and political differences were a source of tension, and each nation vied for territory and influence around the world—even in outer space. This period of rivalry is known as the Cold War, and it heavily influenced geopolitics for the next four decades, including spy games, military overthrows, and political intervention. Both nations began stockpiling nuclear weapons and raced to launch the first satellites (and, eventually, people) into space. While there were other conventional wars fought around the world during this time, the threat of nuclear war is one factor that kept the two superpowers from directly engaging each other in traditional combat.

For both countries, the Cold War meant the further development of computers. By 1950, the Soviet Union had built their first programmable computer, MESM (the Small Electronic Computing Machine). The United States had learned the strategic value of scientific research during World War II and continued to fund computer projects—with no expense spared.

Unlike other nations impacted by World War II, the United States emerged from the conflict with an economy that was still intact. This meant that many Americans became part of a new postwar middle class. Computers were built to support booming businesses, with commercial demand spurring new kinds of innovations. These behemoth computers were out of reach—physically and financially—to the general public. Even so, for the first time, average people became aware of these "thinking machines" and how they could be used to increase their collective "thinking power."

TIMELINE

1951

UNIVAC

UNIVAC (UNIVersal Automatic Computer) was the first commercially successful computer.

1952

FIRST COMPILER

Grace Hopper completed the first implemented compiler, called A-0. Compilers allow programmers to "talk" to computers in words instead of in binary code.

1958

SAGE BECOMES OPERATIONAL

In response to the threat of a surprise Soviet attack, the U.S. military created SAGE (Semi-Automatic Ground Environment), a network of computers to monitor the airspace within and surrounding the United States.

1958

THE FIRST INTEGRATED CIRCUIT

An integrated circuit (IC) has all the components of a computer circuit etched onto a single slice of semiconducting material. ICs allow for more and tinier transistors, making computers both smaller and more powerful.

1965

MOORE'S LAW

As technology improved, transistors continued to get smaller. Intel co-founder Gordon Moore predicted that the number of transistors on a single computer chip would grow exponentially, doubling every two years. His prediction held true for decades, setting a yearly goal for chip manufacturers.

1968

THE MOTHER OF ALL DEMOS

SRI's Douglas Engelbart directly inspired the personal computer revolution when he gave a presentation in San Francisco called "A Research Center for Augmenting Human Intellect." It was a live video teleconference that demonstrated a windows interface, hypertext, graphics, navigation and input, teleconferencing, word processing, and the computer mouse!

FIRST FULLY TRANSISTORIZED COMPUTER

A team at the University of Manchester demonstrated their prototype for the first computer that used only transistors.

1957 FORTRAN DEBUTS

FORTRAN (FORmula TRANslation) is considered one of the earliest high-level computer languages that was widely used. Instead of writing in binary, programmers used a combination of English shorthand and algebraic equations.

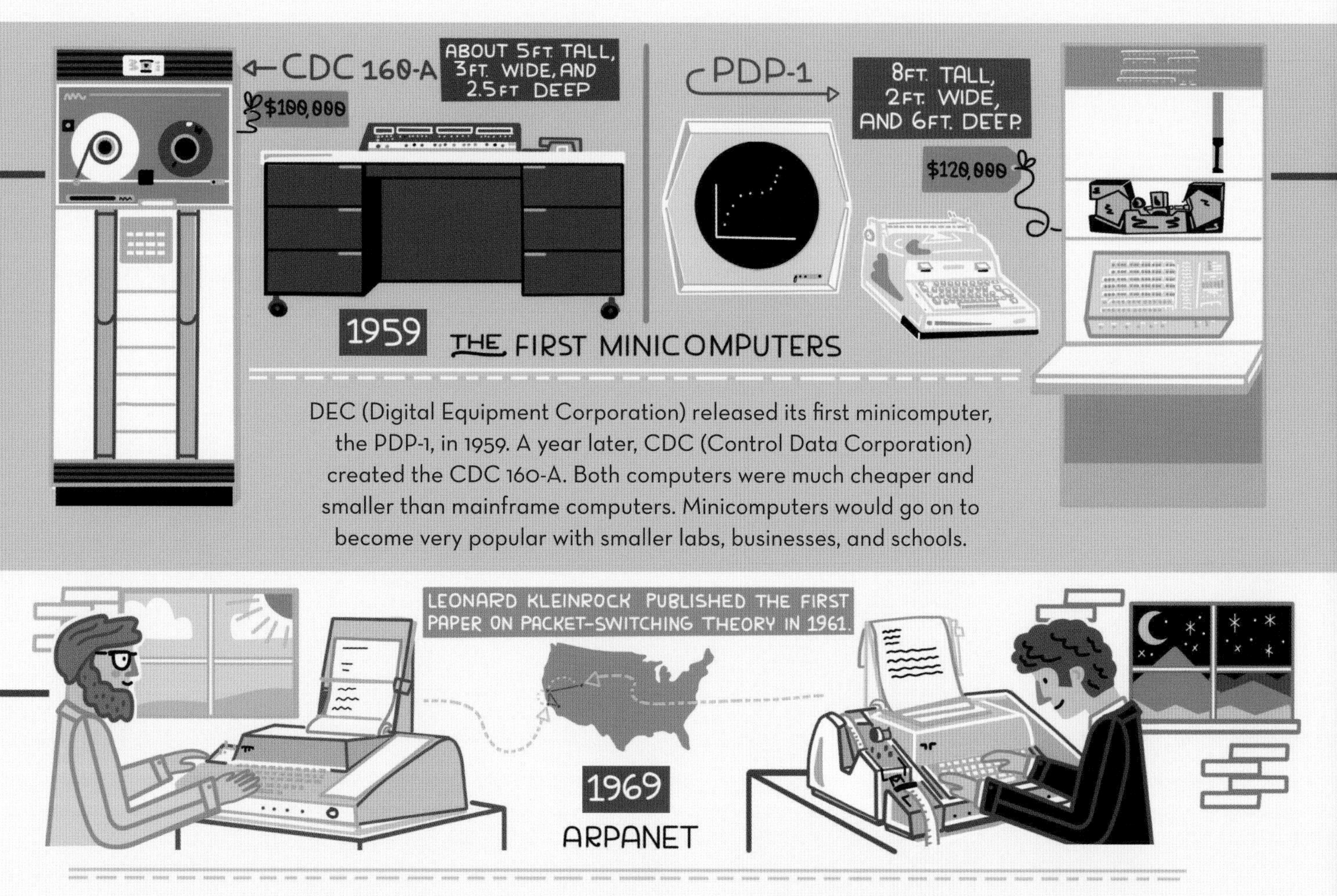

1959 THE FIRST MINICOMPUTERS

DEC (Digital Equipment Corporation) released its first minicomputer, the PDP-1, in 1959. A year later, CDC (Control Data Corporation) created the CDC 160-A. Both computers were much cheaper and smaller than mainframe computers. Minicomputers would go on to become very popular with smaller labs, businesses, and schools.

1969 ARPANET

ARPANET (Advanced Research Projects Agency NETwork), a network of three computers in California and one in Utah, was the very start of the internet. The computers exchanged messages over telephone lines using packet switching. Large messages clogged up telephone lines and made them busy; packet switching broke up messages into smaller data "packets" that each traveled the most efficient path through the labyrinth of phone lines and then reordered themselves on arrival to create the larger complete message.

STORIES FROM HISTORY

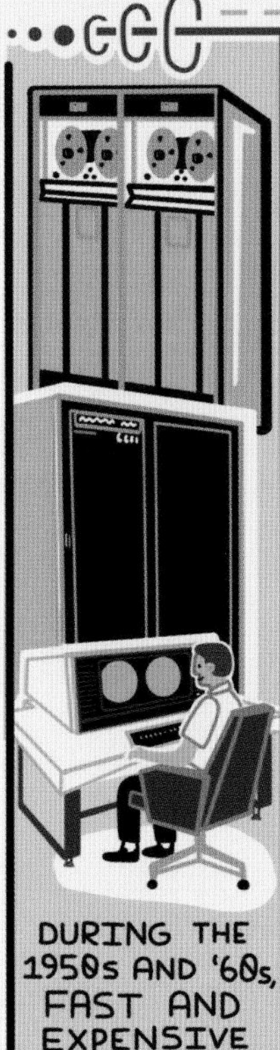

DURING THE 1950s AND '60s, FAST AND EXPENSIVE SUPERCOMPUTERS WERE BUILT BY COMPANIES SUCH AS CDC AND IBM.

THE LARGEST PROGRAM WRITTEN ON PUNCH CARDS WAS FOR THE SAGE PROGRAM.

IT COMPRISED 62,500 PUNCH CARDS (ABOUT 5 MB OF DATA).

Throughout the Cold War, the U.S. military ordered the most expensive science and technology projects in history. The purpose was to build weapons and defense systems, but it also meant that engineers could push the brand-new field of computer science without having to worry about making a profit. Expensive projects such as Whirlwind and SAGE and the creation of NASA (National Aeronautics and Space Administration) were all motivated by the Cold War conflict. These programs indirectly developed technology that we still rely on every day.

WHIRLWIND AND SAGE

In 1945, the U.S. Navy contracted with MIT to build a flight simulator called Whirlwind. The task would be difficult, because they had to invent new technology that could handle the speed, flexibility, and real-time interaction needed for a flight simulator. While building Whirlwind, computer scientist Robert Everett, assisted by Jay Forrester, created the first computer display screen, where the results of programs could be seen using an oscilloscope. Forrester further developed magnetic-core memory, an early type of reliable RAM that could be used for critical programs, like air traffic control. Costing more than a million dollars each year to develop, Whirlwind became operational in 1951 but was never used as a flight simulator. Even so, Whirlwind proved that computers could be used to show maps and visually track objects in real time. Unlike earlier computers, Whirlwind wasn't just a digital replacement for mechanical number crunchers. It inspired radical ideas and showed that computers could interact with real-world objects. The technology gleaned from Whirlwind would be used in the largest and most expensive computer project in history: SAGE.

With the ever-present threat of nuclear attack from the Soviet Union, the U.S. government watched the skies and communicated through a nationwide network of computers. From 1958 through 1984, SAGE monitored air traffic inside and around the United States and controlled NORAD (North American Aerospace Defense Command), an early warning system to an attack. Data was collected from radar towers, patrolling airplanes, and ships and then processed in command centers around the country, which had entire floors filled with mainframe computers. Operators controlled the computers with video consoles and used light guns directly on the screen to select targets. Each operator watched for tiny blips and had to determine if the shape of each small glowing dot was a civilian, ally, or enemy aircraft. SAGE's massive computer network would inspire the development of graphical interfaces and the first version of the internet, ARPANET.

SAGE IS STILL THE LARGEST COMPUTER PROJECT IN HISTORY, COSTING $10 BILLION IN 1954.

THE SPACE RACE

In 1957, the Soviet Union launched the first satellite, *Sputnik 1*, into space. That same year, they also launched a dog named Laika, and, in 1961, their first cosmonaut, Yuri Gagarin. The Soviet space program was considered a huge threat to the United States—how could America be the number-one global superpower if they were number two in outer space? And what if the Soviets found a way to launch missiles from space? To compete in the Space Race, in 1958, the U.S. government created NASA.

Three weeks after Gagarin's space flight, the United States sent their first astronaut, Alan Shepard, into space as part of Project Mercury. Later that year, President John F. Kennedy announced that an American would land on the moon within a decade. The Apollo program was one of the NASA projects dedicated to achieving this goal. It was one of the most massive scientific endeavors in history, requiring NASA to hire more than four hundred thousand people from private industry and academia.

One of NASA's most advanced technological developments was the AGC (Apollo Guidance Computer). On early space missions, like the Mercury project, astronauts flew their spacecraft manually, using control sticks. While astronauts would have loved to pilot the Apollo spacecraft themselves, the distances and complexity involved with flying to the moon and back meant it could only be done by computer. To create a computer small enough to fit in the Apollo command and service module, NASA used cutting-edge technology such as newly invented integrated circuits (see page 60). After nearly a decade of successful space flights, failed missions, and lots of problem solving, in 1969, the Apollo 11 mission succeeded! Astronauts Neil Armstrong and Buzz Aldrin walked on the moon, while Michael Collins flew the command module in orbit, taking in one of the most extraordinary views in human history.

THE FIRST COMPUTER CHIPS WERE USED IN AEROSPACE SYSTEMS.

THE FIRST HEAD-MOUNTED DISPLAY, NAMED THE SWORD OF DAMOCLES, COULD DISPLAY WIREFRAME COMPUTER GRAPHICS IN 1968.

MAINFRAME COMPUTERS GOT THEIR NAMES BECAUSE THE COMPUTERS' ELECTRONICS WERE ATTACHED TO GIANT METAL FRAMES.

POSTWAR CONSUMERISM

While the government was funding computer research projects, the booming postwar economy was ripe to create and sell new business machines. Companies took inspiration and technology from government-funded computer research to transform it for the mass market.

UNIVAC

There were many computer start-ups in the late 1940s, including EMCC (Eckert–Mauchly Computer Corporation). EMCC was founded by J. Presper Eckert and John Mauchly in the wake of the success and acclaim received for their wartime computer, ENIAC. In 1946, Eckert and Mauchly convinced the U.S. Census Bureau to fund UNIVAC to help tabulate data and replace outdated counting machines. Their task of creating a new computer system was monumental, and EMCC was underfunded and had a team of only a dozen engineers. Together, they worked on UNIVAC in a loft above a men's clothing store in downtown Philadelphia. In the summer, the room was so swelteringly hot that the engineers took breaks to dump water on their heads. Despite these conditions, the team furthered the development of memory and storage systems, including magnetic tape. Even though magnetic tape was already used for audio recordings, early computer customers initially mistrusted the "invisible" tape, because it didn't have holes like the punched paper they were used to.

Typewriter company Remington Rand acquired EMCC in 1950, and finally, in 1951, UNIVAC was fully operational and used by the U.S. Census Bureau. UNIVAC became the first commercially successful computer.

BECAUSE OF COST, PROGRAMMING WAS MOSTLY DONE VIA STACKED PUNCH CARDS OR PUNCHED PAPER TAPE.

IT WAS NERVE-RACKING FOR PROGRAMMERS TO KEEP THEIR PUNCH CARDS IN ORDER—A GUST OF WIND KNOCKING THE CARDS AWRY COULD MEAN DAYS OF WORK LOST.

IN 1961, UNIVAC WAS FEATURED ON A *SUPERMAN'S GIRL FRIEND, LOIS LANE* COMIC BOOK COVER.

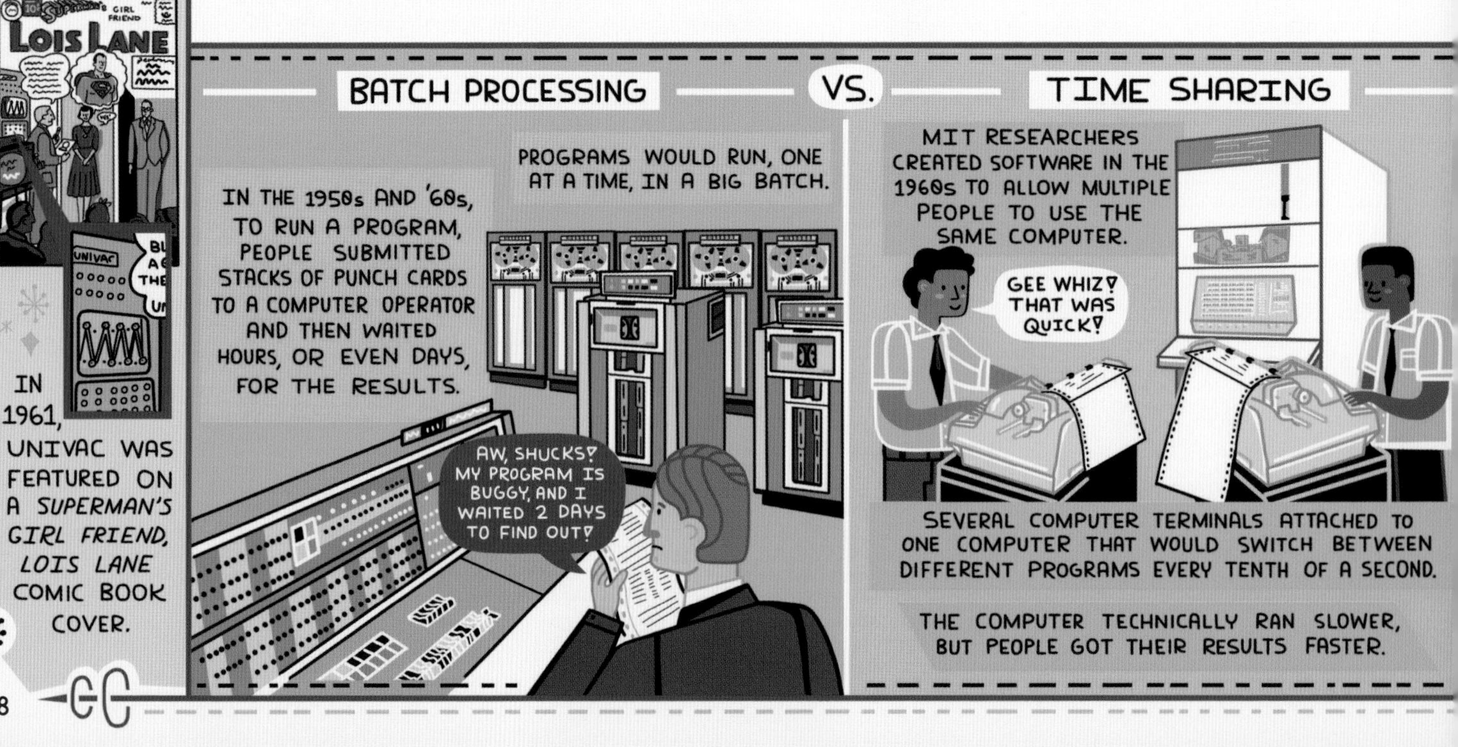

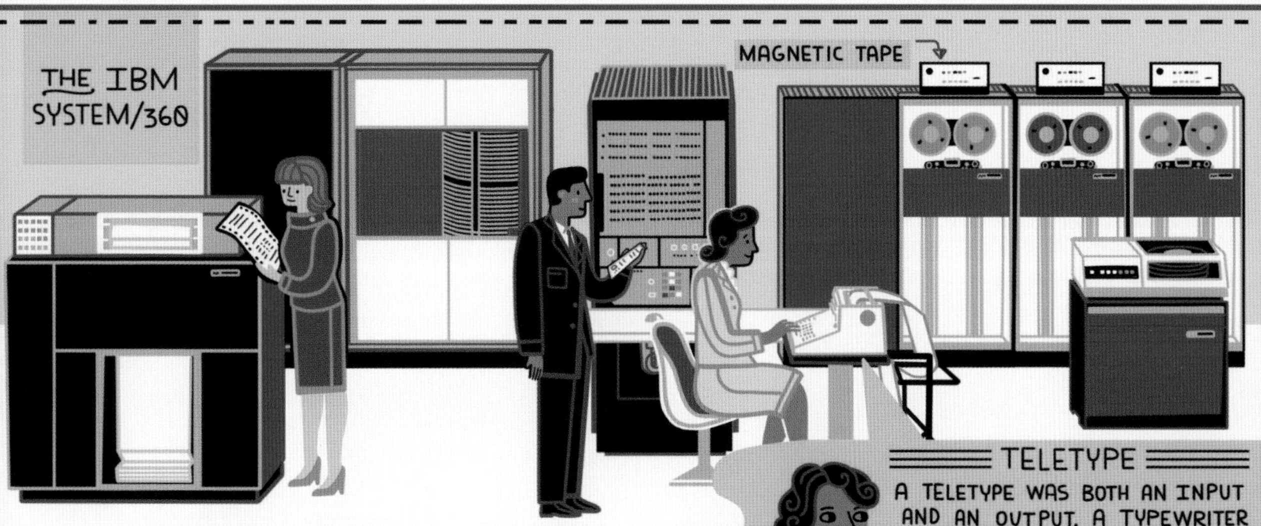

THE IBM SYSTEM/360

MAGNETIC TAPE

TELETYPE

A TELETYPE WAS BOTH AN INPUT AND AN OUTPUT. A TYPEWRITER HOOKED UP TO THE COMPUTER THROUGH TELEPHONE WIRE AND A PAPER PRINTER ACTED AS THE DISPLAY.

FROM THE 1950s TO THE EARLY '70s, TELETYPES WERE HOW MOST PEOPLE ACCESSED COMPUTERS.

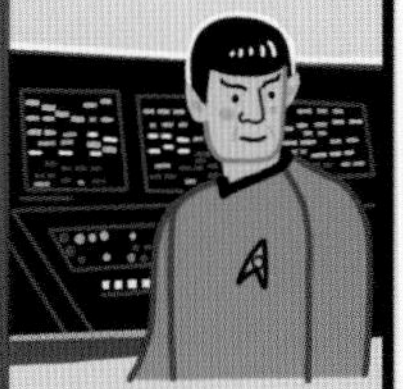

COMPUTERS BEGAN APPEARING ON TV SHOWS LIKE *STAR TREK* (1966) AND IN MOVIES LIKE *2001: A SPACE ODYSSEY* (1968).

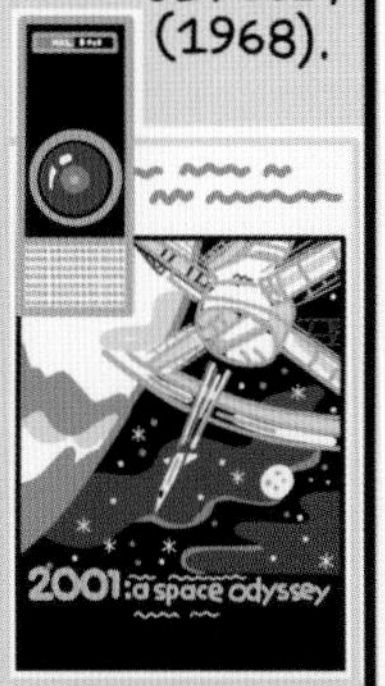

THE FIGHT FOR MAINFRAME DOMINANCE

While EMCC was developing UNIVAC, IBM was focused on government contracts and ignored the commercial market for computers. With UNIVAC's introduction in 1951, it began replacing IBM's outdated office tabulators. This caused IBM to go into panic mode to regain market share. In 1959, IBM introduced their Model 1401 computer system, in a stylish light blue. Its "chain" printer drew sales, hammering out an impressive speed of six hundred lines per minute.

By the 1960s, a third of all computers were made by IBM, but with big success came big problems. IBM had a dozen different computer families and five different product lines, none of which was compatible with the other. It was a mess! In 1959, a secret IBM project called System/360 began working to unify IBM's computers. Delivered in 1965, it was a single computer architecture that shared compatible software on all of IBM's 360 machines. Businesses were no longer locked into one huge computer purchase and instead could upgrade as needed. This "scalability" allowed businesses to buy their first computers, and the IBM 360's popularity helped propel new computer use all over the world.

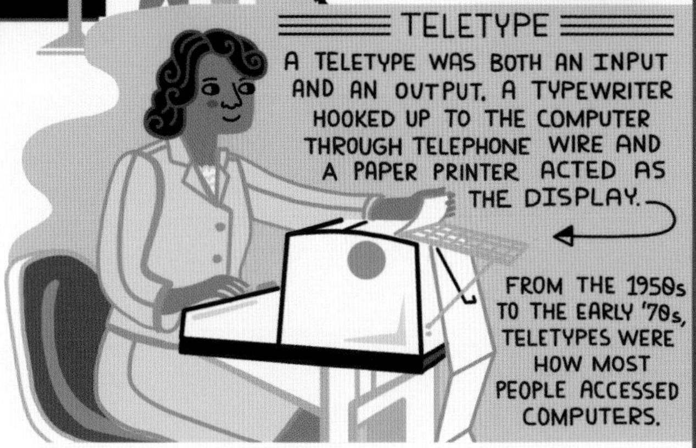

IMPACT OF THE ERA

Two forces worked to further computer science in the 1950s and '60s: the U.S. military, which used its deep pockets to fund big research projects, and the commercial market, which pushed mass production and public awareness of computers. Computers were being used in schools, research labs, and businesses, but they were still far out of reach for the everyday person. Physically massive and expensive, they were handled only by trained specialists in white lab coats. Not even the people programming the computers were allowed to touch the mainframes, which were kept locked away in refrigerated rooms. Despite this, computers became a part of pop culture, and these "electronic brains" with their rows of spinning tape and dials began to inspire literature, movies, and a new generation of "computer nerds" who dreamed of one day having their very own machines.

IBM AND ITS SMALLER COMPETITORS WERE NICKNAMED SNOW WHITE AND THE SEVEN DWARFS: BURROUGHS, HONEYWELL, UNIVAC, NCR, CDC, RCA, AND GE.

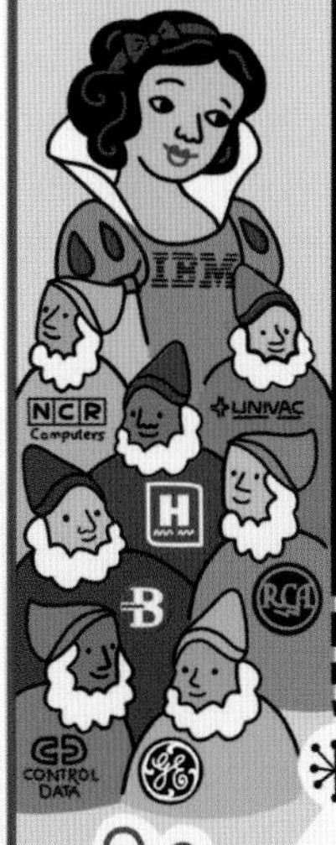

IMPORTANT INVENTIONS

THE FIRST INTEGRATED CIRCUIT A.K.A. THE COMPUTER CHIP—1958

At first, transistors were used in all sorts of devices, from amplifying radios and phones to computers, of course. Computer circuits were built by wiring transistors and other electronics, using tweezers and a steady hand. Computers were bulky and slow, since an electric current had to travel the distances between transistors and separate components. There had to be a better way!

In the 1950s, several scientists and engineering groups worked independently on this problem. While working at Texas Instruments, electrical engineer Jack Kilby realized that an entire circuit could be etched onto a single piece of semiconductor material called germanium. In 1958, Kilby successfully demonstrated his integrated circuit (IC). Meanwhile, the co-founder of Fairchild Semiconductor, Robert Noyce, also invented a different IC, completed in 1959. His was made out of a silicon wafer, had no external wires, and instead used copper connectors. Both Kilby and Noyce have been credited with the invention of the IC. Integrated circuits meant more transistors could be packed into a smaller space. This technology turned computers from room-size machines into pocket-size devices.

7/16 INCH

MADE OF GERMANIUM

GLASS SANDWICHED THE MATERIALS

IN 1961, ROBERT NOYCE RECEIVES THE FIRST IC PATENT.

MADE OF SILICON

SPACEWAR! VIDEO GAME—1962

Fire the torpedoes, it's time to play *Spacewar!* Minicomputers, like the PDP series, were smaller and less expensive than mainframes, making them popular in labs and universities. Inspired by pulp sci-fi novels, Steve Russell and others in the Tech Model Railroad Club designed this video game to show off what a PDP-1 computer could do.

Spacewar! was one of the very first multiplayer video games. Two spaceships faced off, shooting one another. The PDP-1 did more than 90,000 calculations per second to compute user input while making each ship move and fire according to Newtonian physics. *Spacewar!* debuted at MIT in 1962 and went on to inspire the first arcade games.

THE APOLLO GUIDANCE COMPUTER—1966

The AGC was built to navigate the astronauts safely to the moon. It calculated the craft's trajectory based on the astronauts' in-flight measurements of Earth, moon, and star positions and communicated with the spacecraft's guidance systems and many thrusters. NASA needed to build a computer that was small enough to fit inside the Apollo spacecraft and reliable enough to withstand the vibration, radiation, and extreme temperatures of space flight.

The AGC was one of the first computers to use the brand-new (and very expensive) technology of ICs. A 350-person team at the MIT Instrumentation Laboratory, led by software engineer Margaret Hamilton, created the AGC's programming. It used core rope memory, which was nicknamed "little old lady memory" because it was handwoven by female factory workers. Astronauts communicated with the AGC through a numeric display and keyboard called the DSKY (pronounced DIS-kee). In 1968, the AGC successfully navigated the Apollo 8 astronauts to the moon, a major technological feat. While the AGC was one of the most advanced computers of its time, it had about the same computing power as a 1985 Nintendo NES console.

INFLUENTIAL PEOPLE
✦GRACE HOPPER✦1906-1992✦
U.S. NAVY ADMIRAL HOPPER IS KNOWN AS "THE MOTHER OF COMPUTER PROGRAMING."
"THE MOST DAMAGING PHRASE IN THE LANGUAGE IS 'WE'VE ALWAYS DONE IT THIS WAY.'"
SHE JOINED THE NAVY DURING WORLD WAR II AND BECAME SECOND IN COMMAND TO HOWARD AIKEN. SHE WAS ESSENTIAL TO PROGRAMMING THE HARVARD MARK I.
AFTER THE WAR, IN 1949, SHE WAS THE SENIOR MATHEMATICIAN TASKED WITH PROGRAMMING UNIVAC AND LED THE TEAM THAT CREATED THE FIRST COMPILER (1952) AND DEVELOPED PROGRAMMING LANGAUGE FLOW-MATIC (1955).
HOPPER WAS A TECHNICAL ADVISOR ON THE PROJECT THAT CREATED COMPUTER LANGUAGE COBOL IN 1959.
"INNOVATION IS EVERYTHING. WHEN YOU'RE ON THE FOREFRONT, YOU CAN SEE WHAT THE NEXT INNOVATION NEEDS TO BE. WHEN YOU'RE BEHIND, YOU HAVE TO SPEND YOUR ENERGY CATCHING UP."
ROBERT NOYCE 1927-1990
AND
GORDON MOORE 1929-
CO-FOUNDERS OF FAIRCHILD SEMICONDUCTOR IN 1957 AND INTEL CORPORATION IN 1968.
NOYCE IS ONE OF THE INVENTORS OF THE IC.
✦✦IVAN✦✦ SUTHERLAND ✦✦ 1938- ✦✦
"A DISPLAY CONNECTED TO A DIGITAL COMPUTER GIVES US A CHANCE TO GAIN FAMILIARITY WITH CONCEPTS NOT REALIZABLE IN THE PHYSICAL WORLD. IT IS A LOOKING GLASS INTO A MATHEMATICAL WONDERLAND."
HE INVENTED SKETCHPAD IN 1963.
IT WAS ONE OF THE FIRST PROGRAMS TO USE A GUI (GRAPHICAL USER INTERFACE).
USERS DREW ON SCREEN WITH A LIGHT PEN AND COULD ORGANIZE GEOMETRIC SHAPES AND TEXT. IT WAS AN ANCESTOR TO MODERN 3-D DRAWING PROGRAMS. IT COULD EVEN SIMULATE WORKING MACHINES!
SKETCHPAD INFLUENCED DOUGLAS ENGELBART'S NLS AND MANY FUTURE PROGRAMS!
AKIO MORITA 1921-1999 ✦✦✦ AND ✦ MASARU IBUKA 1908-1997
FOUNDERS OF JAPANESE COMPANY SONY
IN THE 1950s, SONY WAS ONE OF THE FIRST COMPANIES TO USE TRANSISTORS FOR NONMILITARY USE.
SONY WOULD BECOME A MAJOR PLAYER IN COMPUTER HISTORY, SETTING THE STANDARDS FOR AUDIO EQUIPMENT, TVs, VISUAL DISPLAYS, AND MUCH MORE!

DOUGLAS ENGELBART 1925-2013

Douglas Engelbart recognized that a computer could become a powerful tool for collaboration. As a young man he read Vannevar Bush's 1945 groundbreaking paper "As We May Think," which inspired him with the idea that if many people had access to a tool that expands mental abilities, then the whole human race could collectively leap forward!

After receiving a PhD in electrical engineering from UC Berkeley, Engelbart worked at SRI (Stanford Research Institute) on the most cutting-edge projects in computing. Funded by NASA and ARPA (Advanced Research Projects Agency), Engelbart led a team of researchers with a broad directive of raising America's collective intelligence. Much like the Apollo program, these large-scale, well-funded science projects had a wide-reaching impact, with some of the tech eventually ending up in the hands of the public.

In the era of "batch processing," when computer users had to wait days for a printout from a program, he envisioned real-time collaboration between users in a graphical environment. Engelbart's team unveiled their years-long project in front of journalists in 1968, later called The Mother of All Demos. They used a giant projection of his computer screen to demonstrate his NLS (oN-Line System). Within this interface of windows and graphics, he called a video conference with a colleague who was miles away and collaborated on a document. They both used a pointer device in the system, the first-ever "mouse." The audience was astonished, realizing they were witnessing a preview of the future.

Although the NLS was never commercialized, many of the people who designed it went on to work for Xerox PARC lab and expanded these ideas. The NLS was designed to augment its user's intellect, while allowing people to work together toward solving humanity's biggest problems. In many ways, Engelbart's vision of interactivity and collaboration was a blueprint that is still being worked toward today.

MARGARET HAMILTON 1936-

The Apollo mission to travel to the moon depended on the work of Margaret Hamilton. After working on SAGE, Hamilton became the head of the Software Engineering Division at the MIT Lincoln Laboratory. At the age of twenty-four, and as a working single mother, Hamilton led the team that created the software for the AGC. Hamilton coined the title "software engineer" and helped define it as a field of study.

The software for the AGC was crucial to navigate the astronauts to the moon. The team also programmed the AGC to perform error detection and recovery in real time. This ended up saving lives. Three minutes before the astronauts were about to touch down on the moon in the Apollo 11 mission, the lunar lander's alarms went off. Red and yellow lights flashed to let the astronauts know the computer was overloaded—running the radar systems and the landing system at the same time required too much processing power. Hamilton and her team had prepared for this possibility, programming the software to prioritize tasks according to importance instead of sequence. The astronauts simply pressed "go" and the computer began the landing sequence with no other problems.

Hamilton went on to work on software for the Skylab space station, and, in 1976, she co-founded Higher Order Software. She was honored with the NASA Exceptional Space Act Award in 2003 and with the Presidential Medal of Freedom in 2016. Hamilton is recognized as the founder of software engineering.

THE ALTAIR 8800, 1974
ALTAIR 8800 COMPUTER
INTEL 4004 MICROPROCESSOR, 1971
C4004
THE WANG 2200 MINICOMPUTER, 1973
WANG
THE APPLE II, 1977
THE XEROX ALTO, 1973
PET
THE COMMODORE PET 2001, 1977

At the end of the 1960s, many people imagined computers to be self-aware and highly advanced, like HAL 9000 in the 1968 film *2001: A Space Odyssey*. In reality, the only relationship an average person had with a computer was as a remote mainframe in a government building, processing their taxes. The advancement of integrated circuit (IC) technology in the 1960s meant that computers could be smaller. Many universities, science labs, and high-end offices had access to "mini" computers. These "small" devices were about the size of a large refrigerator and had many of the same computing abilities of the giant, floor-consuming mainframe computers.

Many young people were inspired by the size of the minicomputers and imagined making an even smaller "micro" computer. What were the possibilities of directing the raw and mysterious power of these machines—originally built for big business and warfare—if they could be placed in your own home? What if you could program your very own computer on a whim? What serious or silly tasks might you make a computer do? These possibilities seemed out of reach throughout the late 1960s, but kept a certain group of teenagers awake at night dreaming of the future. These young tech nerds, inspired by the radical idea of a personal computer, would grow up and go on to create a technology revolution in the 1970s.

TIMELINE

1970 DESKTOP COMPUTERS

Computers such as the Datapoint 2200, which was about the size of a very large typewriter, debuted, but at a price only large governments and businesses could afford. These devices were mostly used to interact with big mainframe computers.

1971 THE UNIX PROGRAMMER'S MANUAL IS PUBLISHED

In 1969, programmers Kenneth Thompson and Dennis Ritchie developed the UNIX operating system while working at Bell Labs. UNIX became popular with engineers and scientists and is the foundation of many operating systems.

1973 COMMUNITY MEMORY

A group of community activists in Berkeley, California, set up computer terminals in coffee shops and record stores in the Bay Area. People were thrilled to be able to send and receive messages through the system and used it as a virtual meeting place.

1977 THE PERSONAL COMPUTER BOOM

Large companies entered the personal-computer market with devices aimed at nontechnical people. Commodore, Apple, and Radio Shack made little computers that worked right out of the box!

THE INTEL 4004 MICROPROCESSOR

This is the first commercially successful microprocessor. Now the "brain" of a computer could be contained entirely inside one small microchip.

THE FIRST MICROPROCESSOR-BASED COMPUTER

In France, François Gernelle designed the Micral N. It was a small computer created for weather-monitoring stations and for controlling water pumps.

GUN FIGHT

Microprocessors used for video games? The Intel 8080 was used to power arcade game *Gun Fight* by Midway.

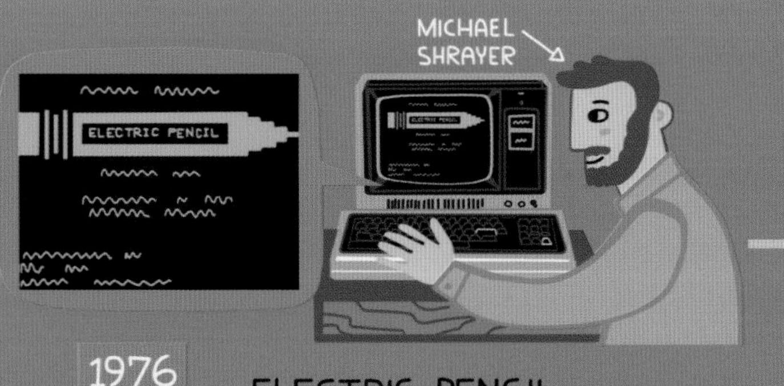

ELECTRIC PENCIL

Typewriters don't have a delete button, but computers do! Computer tinkerer Michael Shrayer created the first word processor for microcomputers, called the Electric Pencil, which let people write books on a computer.

VISICALC

The first spreadsheet software for the Apple II, VisiCalc (visible calculator), brought serious work to the personal computer. Businesses rushed to buy personal computers just to be able to use spreadsheets.

STORIES FROM HISTORY

FROM MINI TO MICRO

In the early 1970s, minicomputers were all the rage with colleges and laboratories. College students and data-entry specialists (who were mostly women) began to come into regular contact with minicomputers as part of their jobs, but they were heavily restricted in the amount of time they had access to the computers. These new minicomputers were exciting and gave many people the inspiration to write programs and think up new uses for computers. Some affluent high schools even offered computer-coding courses on minicomputers. That's how Bill Gates and Paul Allen, the co-founders of Microsoft, discovered computers as teenagers. But at the cost of thousands of dollars, minicomputers were still only for organizations, not for individuals or homes.

THE HOMEBREW COMPUTER CLUB

THE CLUB'S MOTTO WAS "GIVE TO HELP OTHERS."

COOL!

APPLE INC. CO-FOUNDERS STEVE WOZNIAK AND STEVE JOBS WERE AMONG THE MANY PROLIFIC MEMBERS.

LEE FELSENSTEIN

WHACK!

THE CLUB'S NEWSLETTER AND MONTHLY MEETINGS ALLOWED THEM TO EXCHANGE IDEAS, WHICH ENABLED THE PERSONAL COMPUTER REVOLUTION TO HAPPEN.

ITS ORIGINAL FOUNDERS WERE GORDON FRENCH AND FRED MOORE.

THEIR FIRST MEETINGS WERE IN FRENCH'S GARAGE; LATER, THEY MET IN AN UNUSED ROOM AT THE STANFORD UNIVERSITY SCHOOL OF MEDICINE.

Microcomputers, or "personal computers," have a different lineage than big mainframe computers. In the early 1970s, large computer companies, like IBM and HP, could have easily created a small, affordable computer for regular people, if they wanted to. Instead, these billion-dollar companies thought no one would want a computer in their home! In response to this hostile attitude, a rogue group of eccentrics and academics came together near Menlo Park, California, in pursuit of creating some of the first personal computers.

The group, who called themselves the Homebrew Computer Club, attracted a who's who of future computer pioneers. They were far from the white-lab-coat types found on government projects or at IBM—they were tinkerers from ham-radio clubs, teletype hackers, and the nerdiest of the hippies. What this free-spirited group wanted was for the power of computers to become accessible to average people, not just to big corporations.

One of the group's original members was Lee Felsenstein, a "computer liberation" fanatic who had spent years working on Community Memory. Felsenstein often moderated the club's rowdy meetings with a long stick in hand, while club members excitedly shared ideas, spare parts, and designs. Together, the group worked toward creating something that resembled a small computer.

THE UTAH TEAPOT IS A FAMOUS COMPUTER-GRAPHICS MODEL MADE IN 1975. IT'S THE FIRST MODEL TO USE BÉZIER CURVES.

MANY THINK THE ALTAIR 8800 WAS NAMED AFTER A PLANET IN *STAR TREK*. ALTAIR IS ALSO A REAL STAR.

JERRY LAWSON INVENTED THE FIRST CARTRIDGE-BASED VIDEO-GAME SYSTEM, THE FAIRCHILD CHANNEL F (1976).

THE ALTAIR 8800

For a moment in 1975, the center of the computer universe shifted from the Bay Area in California to sleepy Albuquerque, New Mexico. There Ed Roberts, a former U.S. Air Force electrical engineer, created MITS, a company that designed electronics kits for model-rocket hobbyists. The 1960s and '70s were the peak of hobbyist culture. Without cheap global manufacturing, many people built gadgets and electronics from plans available in magazines. Hi-fi stereos, televisions, and even cars were built at home! Microcomputers fit right into that DIY culture.

When Intel discounted some of their then-new 8080 microprocessors, Roberts bought a ton of these tiny computer chips and designed a microcomputer around them. He called it the Altair 8800. It had no keyboard or monitor, just toggles that a user flipped up or down to enter data in binary. Blinking lights showed the results of a program. Although it was a buggy and confusing little box, it was featured on the cover of *Popular Electronics*, and tech nerds went wild for it!

The problem was the Altair couldn't really do anything without an easy-to-use computer language. Harvard University students Paul Allen and Bill Gates understood that if they could take a well-known language, like BASIC (Beginner's All-purpose Symbolic Instruction Code), and get the Altair to run it, there would be endless possibilities for software to run on home computers. Allen convinced Gates to drop out of Harvard and join him. Together they worked on an Intel 8080 simulator and managed to write a version of BASIC small enough to run on the Altair. In March 1975, they scheduled a meeting at MITS to present their software. While en route to New Mexico, Allen realized they had forgotten to write a "boot" program to tell the Altair to load BASIC. He franticly wrote a "boot loader" on scraps of paper during his flight, hoping it would work. At the meeting, their software worked perfectly, and MITS bought it! That same year, Allen and Gates co-founded a software company. Allen suggested the name Micro-Soft, short for microcomputer software.

With Altair BASIC, people were now able to create, on their own little computer gizmos, translations of programs that had originally been written for minicomputers and mainframes. Pretty far out!

THE FIRST COMPUTER WORM WAS "THE CREEPER."

IN 1971, IT HARMLESSLY INFECTED COMPUTERS ON ARPANET TO DISPLAY "I'M THE CREEPER. CATCH ME IF YOU CAN!"

THE FIRST VIDEO ARCADE GAME WAS *COMPUTER SPACE* (1971).

FLOWER

IN 1978, THE SPEAK & SPELL TOY FROM TEXAS INSTRUMENTS USED LINEAR PREDICTIVE CODING TO CREATE A "VOICE."

THE VERY FIRST LOGO FOR APPLE FEATURED SIR ISAAC NEWTON'S DISCOVERY OF GRAVITY.

THE FIRST COMPUTER ANIMATION IN A HOLLYWOOD MOVIE WAS *FUTUREWORLD* (1976), WHICH USED A MODEL OF EDWIN CATMULL'S HAND AND FACE.

STEVEN SASSON INVENTED AN EARLY DIGITAL CAMERA IN 1975.

IT RECORDED PHOTOS ON A CASSETTE TAPE.

THE COMPUTER AS AN APPLIANCE AND APPLE INC.

Dozens of early home-computer companies released kits in the mid- to late 1970s. These kits required hours of delicate work with a soldering gun to put a computer together. Then, after building it, users still had to program it. This was not convenient or simple enough for most people.

In 1975, computer engineer Steve Wozniak was introduced to early personal computers at the Homebrew Computer Club. Fascinated by the new Altair 8800, Wozniak thought he could do it better and designed an elegant all-in-one computer he called the Apple. The Apple I was a complete computer on one silicon board—no assembly required! The user just needed to hook it up to a keyboard and TV. Each time it was turned on, the Apple I loaded its programming language from a cassette tape. This was a huge improvement from the bulky build-it-from-scratch kits. Wozniak and his business-savvy friend, Steve Jobs, co-founded Apple Computer Company in 1976. The Apple I, priced at $666.66 (because Wozniak thought repeating numbers were cool!), was first sold at the Byte Shop, an early home-computer store in Mountain View, California.

The little Apple computer caught the attention of venture capitalist Mike Markkula. In 1977, with Markkula's guidance, Apple received funding to make a sequel, the Apple II. It came in a sleek plastic case and had BASIC preloaded on a ROM chip. This time, the computer was ready to go right out of the box as soon as it was turned on, just like any other home appliance. Apple was not the first company to make an all-in-one home computer, but the Apple II was undoubtedly the most technologically advanced and practical home computer of the 1970s.

THE APPLE I

$666.66

SMALL COMPUTERS BECOME BIG BUSINESS

By 1977, there were several competing personal computers for sale: the Commodore PET 2001, Apple II, and the TRS-80 by Radio Shack. For the first few years, personal computers could do little more than play video games and make noises; most people still thought of them as a short-lived fad. That all changed in 1979, when the Apple II got its first killer app called VisiCalc, a piece of software so compelling that it drove people to buy a computer just to use it. Written by Dan Bricklin and Bob Frankston, VisiCalc was a spreadsheet program that allowed users to input different values into tables and have them change in real time by whatever formula was applied. What would have taken hours on paper was now instantaneous! This spreadsheet program turned personal computers into a serious business overnight, and finance firms rushed to buy them for every desk. This made many small computer companies, like Apple and Commodore, very profitable very quickly.

XEROX PARC AND "THE OFFICE OF THE FUTURE"

IN THE 1970s, PRINTER COMPANY XEROX FORMED PARC (PALO ALTO RESEARCH CENTER).

MANY OF ITS INVENTIONS WERE WAY AHEAD OF THEIR TIME AND WOULD NOT REACH THE PUBLIC UNTIL THE 1980s.

HERE ARE JUST SOME OF THE INVENTIONS THAT CAME OUT OF XEROX PARC.

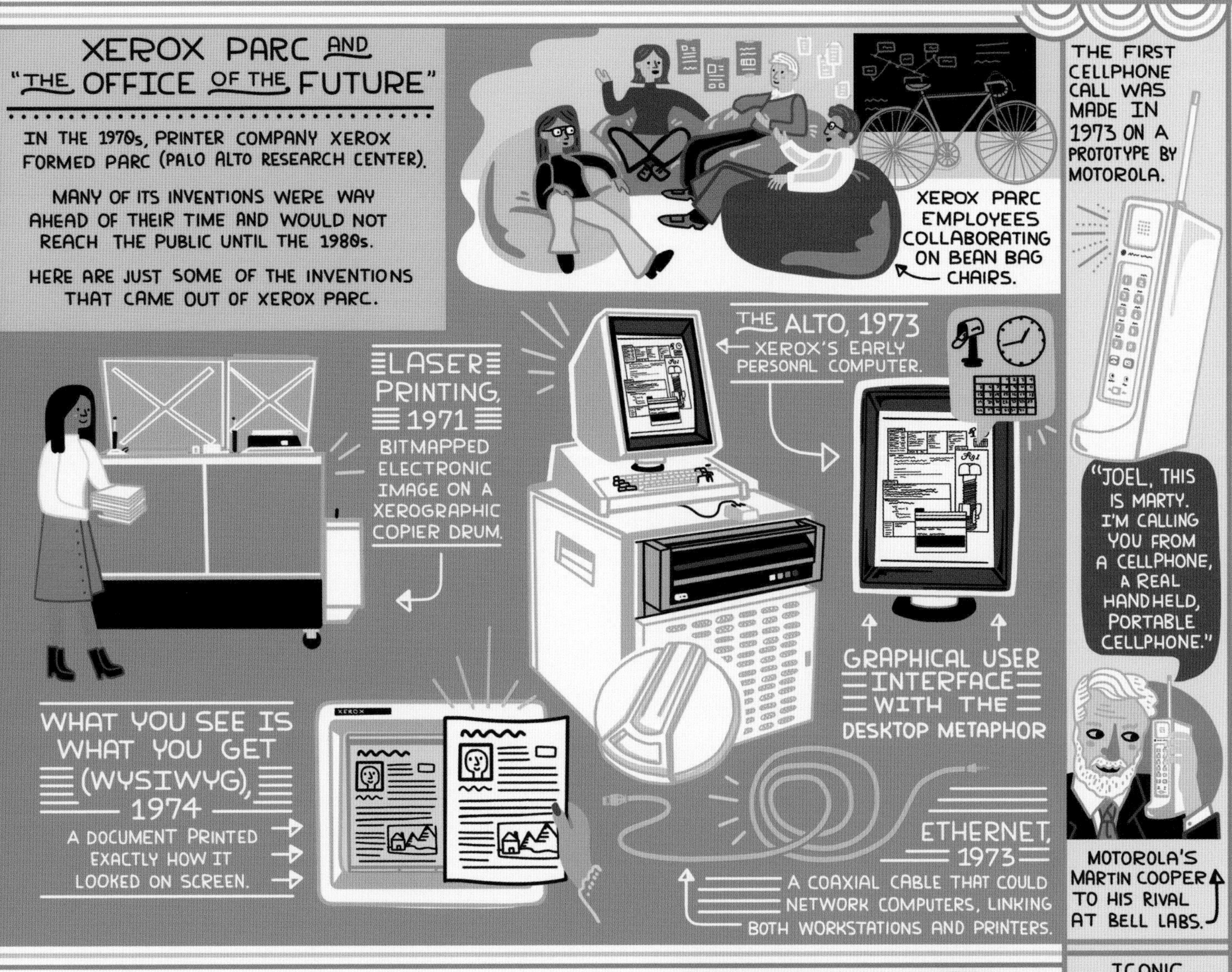

IMPACT OF THE ERA

By the end of the 1970s, the personal computer was an increasingly common fixture in offices, replacing the century-old typewriter. There were many different competing and incompatible personal-computer models on the market. It was a creative and chaotic time, led by many people who did not have any formal business training—but what they did have was a genuine passion for computers. Massive companies, like HP and IBM, were completely absent from personal computers until the 1980s, and that vacuum allowed very futuristic ideas to thrive.

IMPORTANT INVENTIONS

THE MICROPROCESSOR—RELEASED IN 1971

Today, microprocessors are in almost every gadget. In the past, different devices, like a calculator or a timer, each needed their own specially built computer chip, and those chips could do only one specific task. In 1970, Intel hired Federico Faggin to design a single IC chip that could be reprogrammed for different devices, saving time and money. Faggin, with help from Masatoshi Shima on the logic design, devised the architecture of the Intel 4004. They succeeded in creating one chip that could do the job of many different chips—it just had to be programmed with different code.

Microprocessors are general purpose. Operating lawn sprinklers, controlling a pinball machine, or running programming language BASIC—it's all the same to a microprocessor! Although the first microprocessors were not originally intended to be used in computers, enthusiasts hacked them to build the first home computers.

THE INTEL 4004

THE FIRST MICROPROCESSORS WERE BUILT FOR CALCULATORS AND CLOCKS AND WERE USED BY JAPANESE COMPANY BUSICOM FOR THEIR PRINTING CALCULATOR.

THE FLOPPY DISK—1971

Before the floppy disk, programmers had to create software on punch cards, giant rolls of magnetic tape, or hundreds of feet of punched paper, none of which was easy to manage. IBM released the first magnetic floppy disk in 1971 for its word-processing machines, but floppies really took off after they were paired with the Apple II. The disks were small enough to be mailed and could safely carry enough data to fit programs, like VisiCalc, that ran on the new personal computers. This meant that software could be shipped and shared all over the world, and programmers could start companies out of their homes. Floppies continued to be used as a cheap way to distribute software into the 1990s.

FLOPPIES WERE EASY TO PIRATE. IN 1992, THE SOFTWARE PUBLISHERS ASSOCIATION CREATED A PUBLIC SERVICE ANNOUNCEMENT—A RAP SONG, "DON'T COPY THAT FLOPPY!"

DON'T COPY that floppy

8-INCH FLOPPY

5 1/4-INCH FLOPPY

3 1/2-INCH FLOPPY

NETWORKED EMAIL—1971

Before networked email, people could send and receive messages within their time-sharing systems, but these programs were confined to a single computer, and users could share messages only with other people on the system (often only on the same office floor). In 1971, an engineer named Ray Tomlinson created a program called CPYNET to send files between networked computers. He quickly realized he could send messages too. This became the first app on the early internet, which was then known as ARPANET. Two years later, more than 50 percent of the traffic on ARPANET was email!

UGH SPAM!

THE FIRST EMAIL SPAM MESSAGE WAS SENT IN 1978. IT WAS AN AD FOR AN EVENT DEMONSTRATING A NEW LINE OF COMPUTERS. MORE THAN ONE HUNDRED ARPANET USERS GOT THE EMAIL, MOST OF WHOM WERE ANNOYED.

INFLUENTIAL PEOPLE
"THE BEST WAY TO PREDICT THE FUTURE IS TO INVENT IT."
WORKED AT XEROX PARC AND HELPED MAP OUT THE FUTURE OF PERSONAL COMPUTERS.
HIS WORK DIRCTLY INFLUENCED THE FUTURE DESIGN OF ALL TABLET DEVICES.
KAY HAS FOCUSED ON CLIMATE CHANGE AND WORKS WITH THE ELLEN MACARTHUR FOUNDATION.
• ALAN KAY • 1940– •
KAY CREATED THE CONCEPT FOR THE DYNABOOK, A SMALL, PORTABLE COMPUTER FOR TEACHING CHILDREN.
MICROSOFT
HE CO-FOUNDED MICROSOFT WITH BILL GATES.
WROTE BASIC FOR THE ALTAIR 8800 IN 1975.
"LANGUAGES EVOLVE; IDEAS BLEND TOGETHER. IN COMPUTER TECHNOLOGY, WE ALL STAND ON OTHERS' SHOULDERS."
HE WAS A GUITAR VIRTUOSO, YACHT CAPTAIN, MUSEUM CURATOR, AND AN NFL AND NBA TEAM OWNER!
•PAUL ALLEN•1953-2018•
AFTER NEGOTIATING A DEAL TO HAVE MICROSOFT'S SOFTWARE ON EVERY IBM PC IN THE 1980s, ALLEN LEFT THE COMPUTER WORLD AND PURSUED HIS MANY INTERESTS.
"SUCCESS IS MORE A FUNCTION OF CONSISTENT COMMON SENSE THAN IT IS OF GENIUS."
•AN WANG 1920–1990•
HE INVENTED CORE MEMORY, USED IN THE MAINFRAME COMPUTERS OF THE 1960s.
WANG
BORN IN CHINA, HE IMMIGRATED TO THE U.S. IN 1945 TO EARN A PHD IN PHYSICS AND ENGINEERING FROM HARVARD.
FOUNDED WANG LABORATORIES IN 1951. WANG LABS CREATED SOME OF THE MOST SUCCESSFUL WORD PROCESSORS, CALCULATORS, AND OFFICE COMPUTERS OF THE 1970s AND '80s.
•LEE FELSENSTEIN•1945–•
HE HELPED DEVELOP THE COMMUNITY MEMORY PROJECT AND WAS A KEY MEMBER OF THE HOMEBREW COMPUTER CLUB.
"TO CHANGE THE RULES, CHANGE THE TOOLS."
DESIGNED THE PENNYWHISTLE MODEM AND HELPED DESIGN AN EARLY MICROCOMPUTER, THE SOL-20 (1976).
HE CO-FOUNDED THE OSBORNE COMPUTER COMPANY WITH ADAM OSBORNE AND DESIGNED THE OSBORNE 1 (1981). IT'S CONSIDERED THE FIRST PORTABLE COMPUTER—AT 24.5 LBS!
FEDERICO FAGGIN •••1941–•••
BORN IN ITALY, HE MOVED TO THE U.S. TO WORK FOR FAIRCHILD SEMICONDUCTOR IN 1968. TWO YEARS LATER, HE INVENTED THE FIRST MICROPROCESSOR FOR INTEL.
IN 1974, HE CO-FOUNDED ZILOG, INC. AND CREATED THE Z80 MICROPROCESSOR (ONE OF THE MOST SUCCESSFUL 8-BIT MICROPROCESSORS EVER).
HE CO-FOUNDED SYNAPTICS INC., WHICH CREATED THE MODERN TOUCHPAD IN 1986.

"[THE INTEL 4004] WAS A VERY PRIMITIVE COMPUTER BY ANYONE'S STANDARD, BUT IT FORETOLD THE POSSIBILITY OF ONE'S OWN PERSONAL COMPUTER THAT NEED NOT BE SHARED BY ANYONE ELSE."

GARY KILDALL 1942–1994

In 1971, when Intel designed the first microprocessor, the 4004, the tech world thought it could be used only in calculators or industrial machines. Gary Kildall, a young mathematics professor in Monterey, California, was introduced to the first microprocessor while working part-time at Intel. Kildall recognized that these tiny complex chips had the incredible potential to use high-level computer languages instead of only being programmed in machine code.

While growing up, Kildall heard his father, a professional sailor, describe an idea for a mechanical device called the crank, which could calculate a ship's position anywhere in the world, almost like a small computer. When Kildall became fascinated by the Intel 4004, the idea of his father's "crank" inspired him to push the capabilities of the tiny central processing unit. He spent many late nights (even sleeping in a van outside Intel's office) translating a computer language that the Intel 4004 could understand. His friends later said he mostly did it for fun, since it was a nearly impossible task to get a language designed for mainframe computers to work with a microchip meant for a digital wristwatch.

In 1973, Kildall successfully finished PL/M (Programming Language for Microcomputers), the first high-level language for microprocessors that ran on the 8-bit Intel 8008. He followed this with an operating system, CP/M (Control Program for Microcomputers), and he and his wife, Dorothy, founded Digital Research Inc. His foundational work, building a computer system around a microprocessor, is the form that all future computers would take.

STEVE WOZNIAK 1950–

Steve "Woz" Wozniak, co-founder of Apple, started out by building computers for fun! He began his tech career by pulling pranks. As an engineering student, he created a TV jammer out of radio parts that he used to secretly torment the faculty. As a teenager, Wozniak read technical manuals for the minicomputers of the 1960s. He became fascinated with the idea of building a computer out of the fewest available parts and sketched out his designs on paper, dreaming of a time when he would be able to afford them. After college, Wozniak got a job at HP in California, designing calculators. In his free time, he hung out with his friend Steve Jobs, a fellow electronics enthusiast and prankster.

In 1971, the two Steves started their first business selling Blue Boxes, which were small phone-hacking devices that let users place free long-distance calls. These boxes were not exactly legal, but they were really fun! Wozniak and Jobs even tried to prank the Pope when they used a Blue Box to call the Vatican. In 1975, Wozniak began designing the Apple I computer. He offered the design to HP, but they were not interested. Jobs understood the sales potential of a personal computer, and he and Wozniak founded Apple Computer Company in 1976.

Wozniak loved arcade games and had designed the circuit boards for a hit Atari game, *Breakout*, in 1975. When he built the Apple II in 1977, many groundbreaking features were put in specifically to play this game. That meant circuitry for color graphics and sounds and a game paddle. Wozniak continued to work at Apple into the mid-1980s. Since then, he has taken the time to work on a variety of his own electronics projects, teach elementary school, and become a spokesperson for education and new technology.

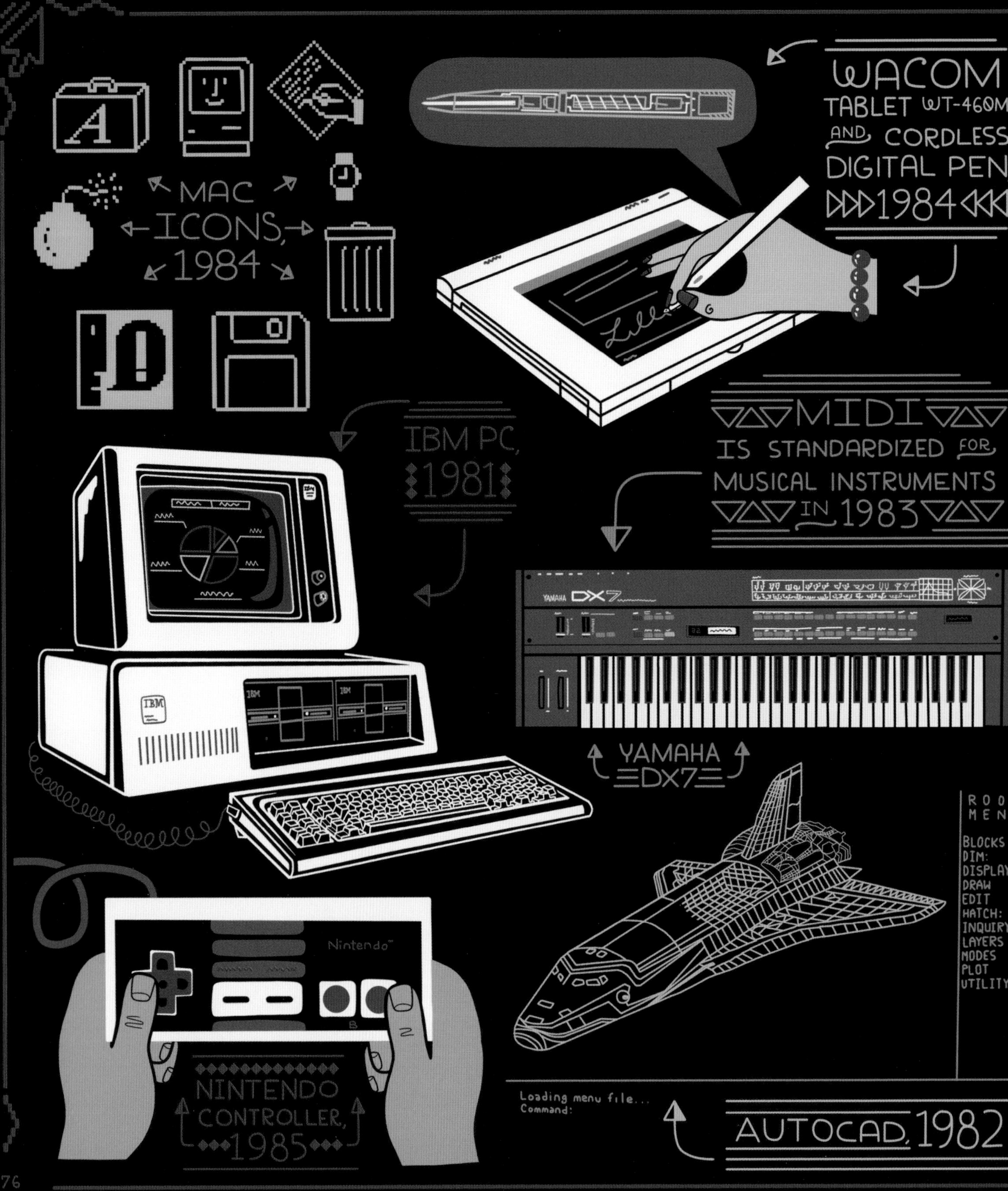

WACOM
TABLET WT-460M
AND CORDLESS
DIGITAL PEN
1984
MAC
ICONS,
1984
IBM PC,
1981
IBM
MIDI
IS STANDARDIZED FOR
MUSICAL INSTRUMENTS
IN 1983
YAMAHA DX7
YAMAHA
DX7
Nintendo™
NINTENDO
CONTROLLER,
1985
BLOCKS
DIM:
DRAW
EDIT
HATCH:
LAYERS
MODES
PLOT
Loading menu file...
Command:
AUTOCAD, 1982

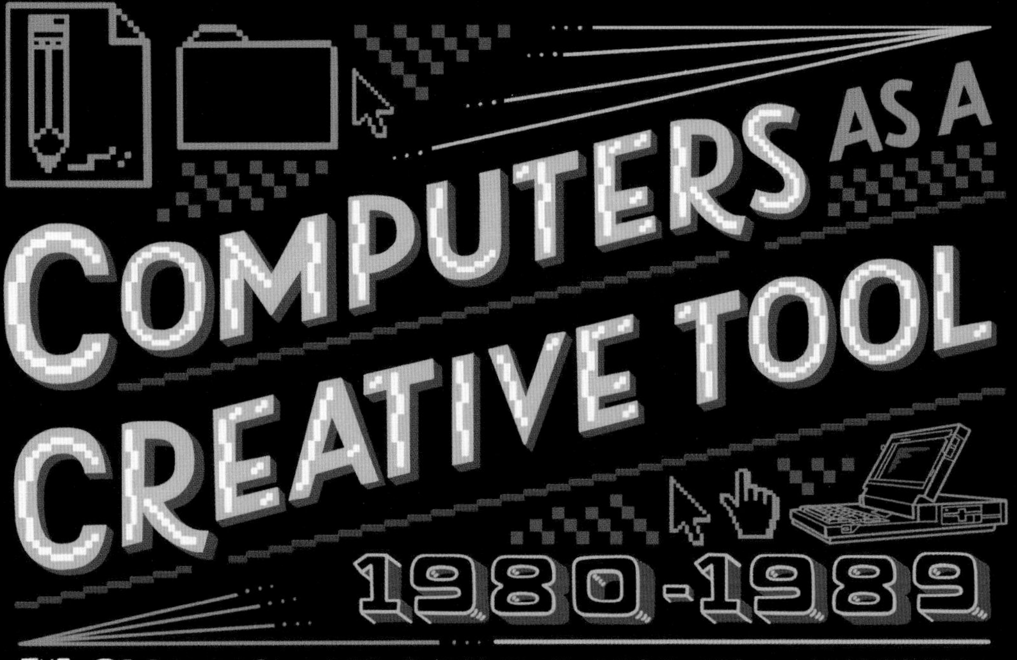

THE GRAPHICAL USER INTERFACE BECOMES MAINSTREAM

In the 1970's, the personal-computer revolution made it possible for people to finally afford (and physically fit) a computer in the home. Computers were still a technologically advanced device that required a pretty steep learning curve. Users had to type in complicated commands to do anything. Another radical shift would have to happen before the computer was accessible for all people: the GUI (graphical user interface).

GUIs transformed a computer's dark glowing screen filled with lines of ghostly green text into a "desktop" with easy to understand, recognizable icons. By using a mouse to point and click on visual icons, anyone could use a computer. In the 1980s, the visual designs for GUIs were further developed and commercialized, and more and more people began buying computers. A suit-wearing office worker might have had an IBM PC to bring work home. Graphic designers were enthralled with the stylish Macintosh. And young people (or anyone else without deep pockets) could snag a colorful and inexpensive Commodore 64. The technology advanced so quickly in the 1980s that laptops were even used on the space shuttle.

The personal-computer boom further helped turn these machines into creative tools. Special software enabled new kinds of music, movies, and artwork to be created. It is amazing to think that just a few decades earlier, computers had been built exclusively for the purpose of crunching numbers and calculating missile trajectories—and now they were being used by artists! In the 1980s, the computer had officially matured into an essential tool for creative professionals.

TIMELINE

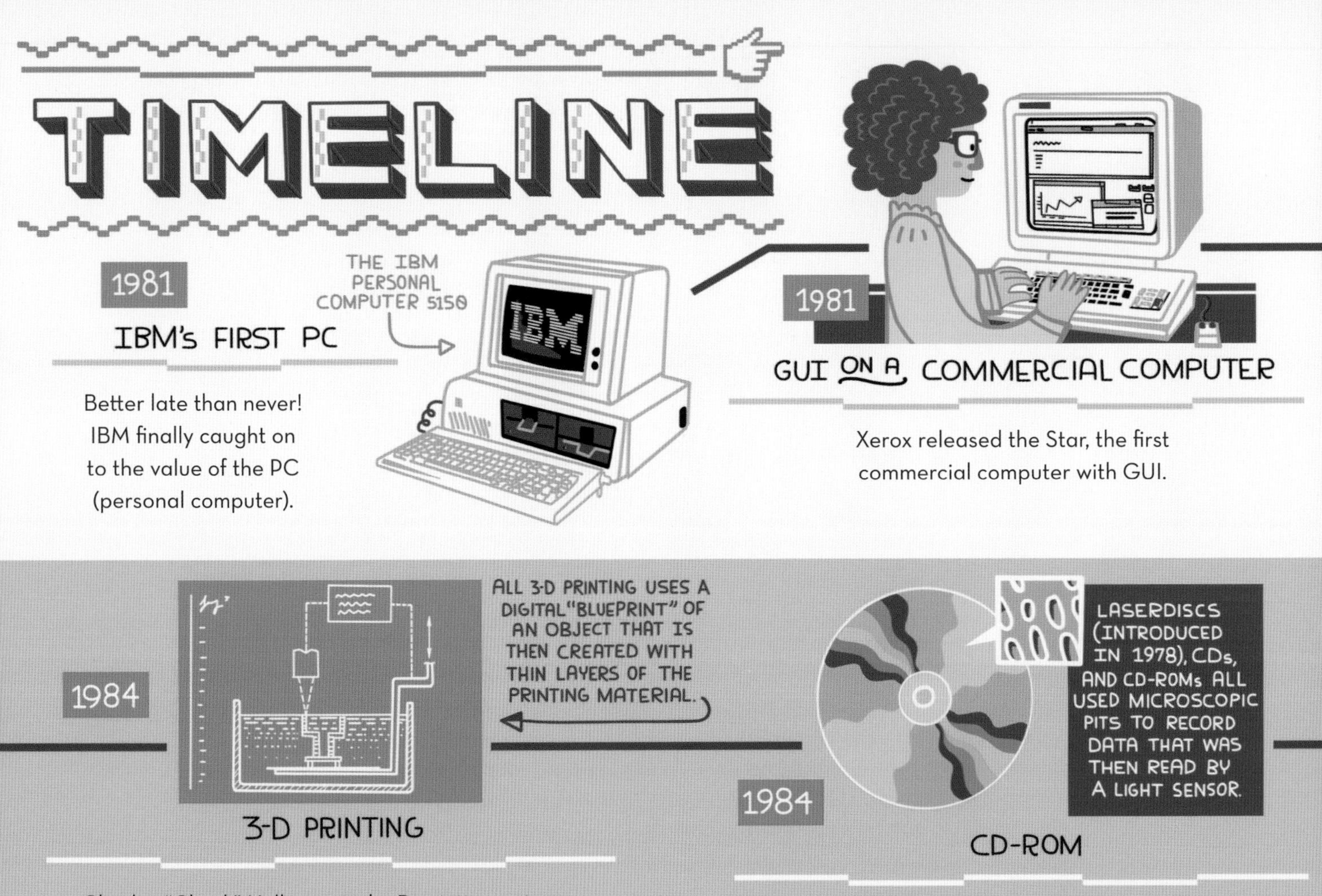

1981

IBM's FIRST PC

Better late than never! IBM finally caught on to the value of the PC (personal computer).

1981

GUI ON A COMMERCIAL COMPUTER

Xerox released the Star, the first commercial computer with GUI.

1984

3-D PRINTING

Charles "Chuck" Hull invented 3-D printing, using polymers that harden under UV (ultraviolet) light. By controlling the position of a UV laser with a computer, he could build a 3-D shape, one layer at a time.

1984

CD-ROM

The CD (compact disc) was introduced for digital audio in 1982. The CD-ROM allowed all sorts of media—software, video games, and books—to be stored on CDs and distributed at a low cost.

1986

PIXAR

Pixar Animation Studios, which is known for its computer-animated films, was formed when Steve Jobs bought the special-effects Graphics Group from Lucasfilm.

NINTENDO 8-BIT VIDEO-GAME CONSOLE

Nintendo released its home entertainment system called Family Computer (better known as the Famicom) in Japan in 1983. Two years later, it was remodeled and released in the United States as the NES (Nintendo Entertainment System). The NES completely revitalized the video-game industry in America.

FLASH MEMORY

While working at Toshiba, Japanese computer-scientist Fujio Masuoka invented flash memory. In 1984, he presented a paper on his memory design at the International Electron Devices Meeting. Flash memory is a nonvolatile memory chip that can be erased and reprogrammed many times.

NSFNET IS FOUNDED

The National Science Foundation created NSFNET (National Science Foundation NETwork) by linking five supercomputer centers located at universities across the United States. Parts of ARPANET and smaller university networks soon began to join the network. NSFNET eventually became the backbone of the internet.

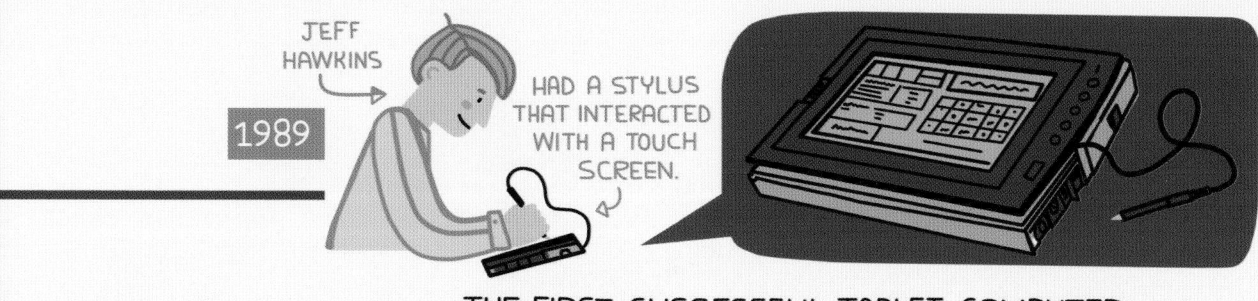

THE FIRST SUCCESSFUL TABLET COMPUTER

Released by GRiD Systems Corporation, the GRidPad 1900 was expensive, heavy (4.5 lb), and used primarily by the U.S. military. Its principal architect and designer was Jeff Hawkins, who went on to create the PalmPilot.

TIME MAGAZINE NAMED THE PERSONAL COMPUTER THE "MACHINE OF THE YEAR" INSTEAD OF A "MAN OF THE YEAR" FOR THE JANUARY 1983 COVER.

IN THE 1980s, JAPANESE COMPANIES WERE LEADERS OF THE SEMICONDUCTOR INDUSTRY!

IN 1981, THE IBM PC WAS RELEASED WITH *MICROSOFT ADVENTURE*, AN EARLY FANTASY ROLE-PLAYING GAME THAT DEEPLY CONFUSED IBM'S SALESMEN.

STORIES FROM HISTORY

By 1980, there were many small personal-computer companies, although only a few had enjoyed any meaningful success. Many of these start-ups had tried unsuccessfully to convince other companies to purchase their buggy and often incompatible computers. Most businesses were waiting for IBM to enter the personal-computer market, since they felt safer spending thousands of dollars on new tech from a reliable and familiar company.

MICROSOFT AND THE PC CLONES

IBM had been caught completely off guard by the PC revolution. They were so far behind that they had to assign a team, outside of their typical years-long development schedule, to create a personal computer. Meanwhile, Microsoft was already the successful publisher of BASIC for every major home computer. IBM approached Microsoft looking for a DOS (disc operating system) for their PC and, with some scrambling, Microsoft managed to slap together PC DOS and license it to IBM—all while keeping the rights to sell it to IBM's competitors!

Engineer Don Estridge led the IBM PC team with input from Bill Gates and Microsoft. Twelve computer engineers (nicknamed the "dirty dozen") flew to Boca Raton, Florida, to build the computer away from IBM bureaucracy. In a surprise move, IBM developed a PC with an open expandable system that allowed anyone to design new components and peripherals. It was full of expansion slots, just like the earlier Altair or Apple II. It was unlike anything IBM had ever made before. The corporate suits at IBM had made a computer directly influenced by hobbyists and tinkerers from the Homebrew Computer Club!

The IBM PC was released in 1981, and within two years it had replaced the Apple II for business use. But there was a catch. To save time, instead of using only IBM tech, they used off-the-shelf computer parts and preexisting software. This meant that other manufacturers could build IBM "clones" from the same parts and then buy a license for MS-DOS from Microsoft, allowing them to sell comparable machines at cheaper prices. The business world completely got behind these IBM compatibles, which were simply called PCs. Within a few years, IBM would be squeezed out of its own market by cheaper clone brands, like Compaq and Dell. Although *PC* stands for "personal computer," the term began to refer only to these many clones. No matter the manufacturer, a PC would contain an Intel compatible microprocessor and a Microsoft operating system. This combination of parts and software dominated personal computers for decades.

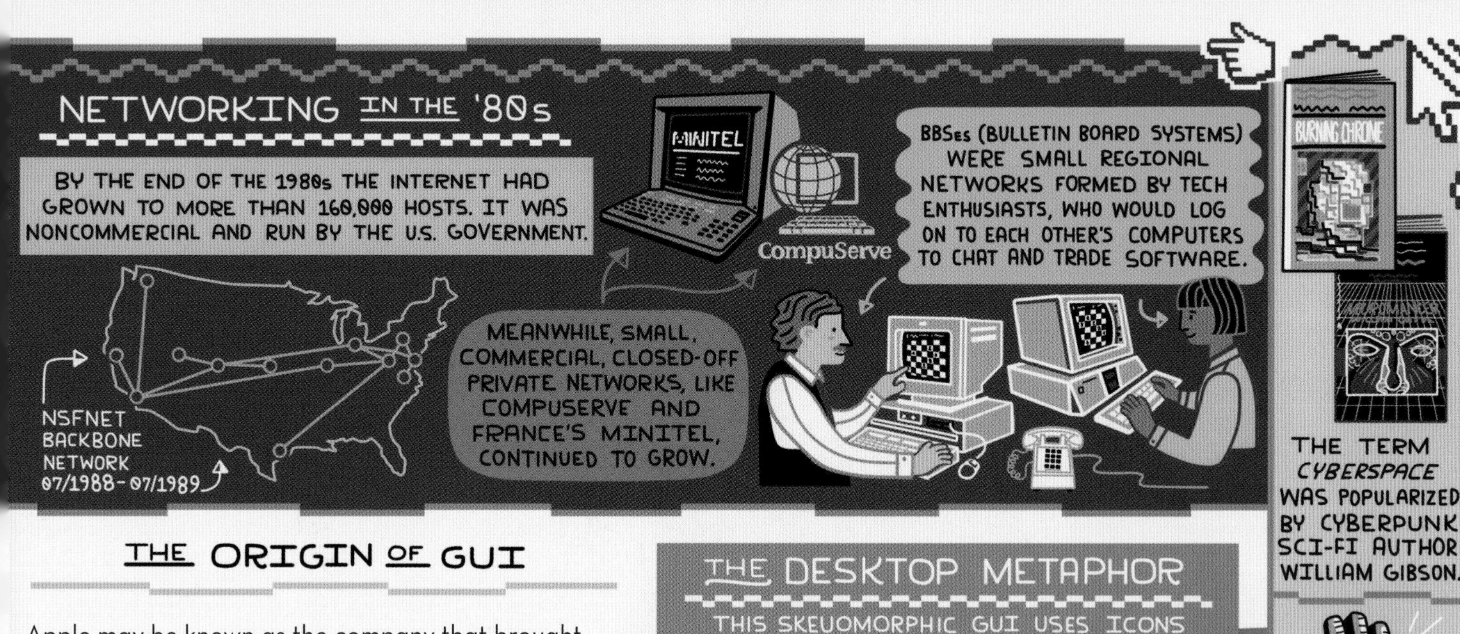

THE ORIGIN OF GUI

Apple may be known as the company that brought "cool" to computers—the Macintosh, released in 1984, is iconic for its design and function, thanks to its sleek and intuitive GUI—but many of the features of the Macintosh were pioneered years earlier at photocopier manufacturer Xerox's PARC lab in Palo Alto, California. This experimental group employed some of the brightest minds in computer engineering. Throughout the 1970s, researchers at PARC invented all sorts of important technology, like laser printing, the desktop GUI, Ethernet networking . . . the list goes on!

In 1973, PARC created an advanced minicomputer called the Alto, which featured a mouse and clickable computer files within windows, just like Douglas Engelbart's 1968 NLS. The Alto's GUI used bitmapped graphics, much like an arcade game, instead of primitive text. This enabled a user to simply click on an icon using a mouse instead of typing commands.

The icons on the Alto's screen were made to resemble their real-world counterparts, in what is known as skeuomorphic design. This meant an icon for an email file looked like an envelope, the icon to tell time looked like a clock, and so on. This was a huge breakthrough and allowed people completely unfamiliar with computers to learn how to use them intuitively.

THE DESKTOP METAPHOR

THIS SKEUOMORPHIC GUI USES ICONS THAT REPRESENT A REAL-WORLD DESKTOP.

LisaWrite Paper

WasteBasket

ClipBoard

GUI ON THE APPLE LISA (1983)

The Alto became well-known in Silicon Valley. Steve Jobs and engineers at Apple visited Xerox PARC in 1979 and were impressed by the windowed graphics. Many PARC employees ended up working at Apple during the 1980s. The exchange of ideas between Xerox and Apple helped further develop GUIs for the consumer market.

THE POWER GLOVE WAS AN EARLY WAY TO CONTROL VIDEO GAMES WITH GESTURES. RELEASED IN 1989, IT WORKED POORLY.

HEATHKIT RELEASED THE HERO JR. (RT-1) "HOME ROBOT" KIT IN 1984. IT WAS PROGRAMMED TO "REMAIN NEAR HUMAN COMPANIONS" BY USING SONAR AND LISTENING FOR VOICES.

THE 1984 SUPERBOWL TV COMMERCIAL INTRODUCING APPLE'S MACINTOSH WAS BASED ON GEORGE ORWELL'S DYSTOPIAN NOVEL *1984*.

STEVE JOBS FOUNDED COMPUTER COMPANY NeXT, INC. IN 1985.

THE FIRST MAC

By the mid-1980s, microprocessors were powerful enough to support GUIs. In 1983, Apple released the Lisa, one of the early consumer computers with a GUI. While the Lisa was attention-grabbing and expensive (it cost as much as a new car), it couldn't compete in a market that was dominated by the relatively affordable IBM PC. Apple co-founder Steve Jobs was undeterred—he was obsessed with beating IBM by bringing futuristic tech to market. He put all of his attention on Apple's next computer, the Macintosh, and pushed the small team working on it to exhaustion.

The Mac, as it came to be known, was designed to be affordable for small businesses. It used a mouse and an intuitive (and fun-looking) desktop GUI. Its specialty was the wide variety of text fonts that could be displayed on its high-resolution, grayscale 9-inch screen. This was groundbreaking in a PC world of green or amber text on dark screens! The Mac was built around the What You See Is What You Get (WYSIWYG) system, first developed at Xerox, meaning that whatever was printed on paper would be identical to what was displayed on the screen. This seems simple now, but back in the 1980s, printing directly from a computer was a tedious mess.

When the Mac was released in 1984, despite all of its new features, it sold so poorly that it was considered a borderline failure. Jobs's tunnel vision in trying to sideline the profitable Apple II with the Mac led to Apple losing money in 1985—a first for the company.

Luckily, several technologies arrived in the nick of time to save the Mac, turning it into a tool for desktop publishing. In 1985, Apple released the LaserWriter, one of the first desktop laser printers, and AppleTalk, a protocol and hardware for a LAN (local area network). This meant a business could replace expensive drafting and publishing equipment by simply networking several Macs to a laser printer. This led to a boom in desktop publishing and considerably lowered the bar for entry into graphic design and print media.

The Mac was first considered an "artsy" machine with a small market, but it proved that the computer would become an essential tool for creatives. It was further developed, and the Mac line of computers went on to become the industry standard for artists.

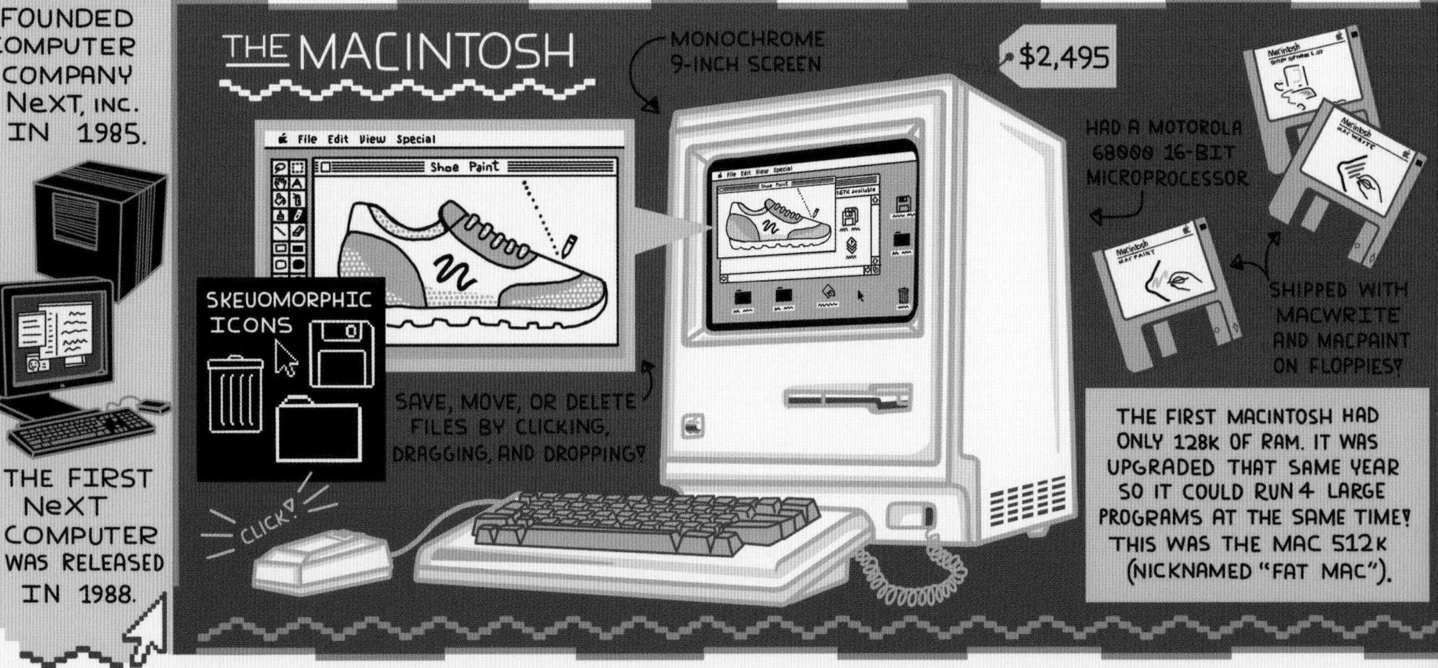

THE FIRST NeXT COMPUTER WAS RELEASED IN 1988.

CGI AND MOVIES

Computer animation developed out of the need to display information, like graphs or drawings. In the early 1960s at Bell Labs, artists would painstakingly photograph each frame of an oscilloscope drawing to create a vector animation. Years later, animators would use primitive wireframe CAD (computer-aided design) programs to create animations for movies such as Lucasfilm's *Star Wars* (1977) and Disney's *The Black Hole* (1979).

Modern CGI (computer-generated imagery) movies began at Pixar Animation Studios. In 1986, Lucasfilm sold its Graphics Group to Steve Jobs, who then formed Pixar with co-founders Edwin Catmull and Alvy Ray Smith. Catmull and Smith had computer engineering backgrounds and had worked at Lucasfilm to create some of the first color CGI sequences in commercial films. Smith had worked at Xerox PARC in the 1970s on SuperPaint, the first computer "paint" program.

With Steve Jobs's vision of using the computer as an artistic tool, Pixar flourished and vastly improved many of the techniques vital to CGI, such as shading, lighting, and particle simulation. By 1989, CGI graphics could be used to create photorealistic scenes. Pixar created RenderMan, the industry-standard software used in its 1988 Oscar-winning short film *Tin Toy*. Pixar would go on to create the first feature-length CGI film, *Toy Story*, in 1995, and CGI technology soon became so advanced that it was difficult to tell if an image was real or created on a computer.

TIN TOY (1988)

WIREFRAME TEST FROM PIXAR'S *LUXO JR.* (1986).

IMPACT OF THE ERA

By the end of the 1980s, the personal computer was no longer an esoteric, number-crunching box that required specialized knowledge to operate. The combination of an affordable price point, an intuitive GUI, and new kinds of software turned computers into a tool for all. Schools and offices could now afford to purchase multiple computers. It became normal for people to use computers to take office work home, play games, or even make music. People were no longer intimidated to boot up and start clicking around!

COOL BEAT!

CHECK OUT MY 'ZINE!

SUPER FUN!

MICROSOFT WORD, ORIGINALLY CALLED MULTI-TOOL WORD, WAS RELEASED IN 1983.

BY 1989, IT WAS THE GLOBAL STANDARD FOR WORD PROCESSING.

A CRAY SUPERCOMPUTER WAS USED TO CREATE THE GRAPHICS IN *TRON* (1982).

TRON

TETRIS

IN 1989, NINTENDO RELEASED THE GAME BOY, A POPULAR HANDHELD VIDEO-GAME CONSOLE.

IMPORTANT INVENTIONS

COMPUTERS FOR THE MASSES: THE COMMODORE 64 – 1982

The little computer that could! The Commodore 64 (C64) was the bestselling computer model of the twentieth century by a wide margin, selling at least twelve million units over its twelve-year life. With a "massive" 64 KB of memory and color graphics, the C64 was a bargain in the early 1980s. Thanks to Commodore founder Jack Tramiel's relentless pursuit of cost reduction, its price tag was half that of its competitors. Marketed as a video-game platform and widely available in toy stores, the C64 gave many people their first programming experience, especially kids!

THE FIRST LAPTOP – 1982

The GRiD (Graphical Retrieval Information Display) Compass was not the first portable computer but its patented "clamshell" design made it the first true laptop! It had many other features that we still see on modern laptops today. Its magnesium metal case was so sturdy that a Compass was taken into space aboard NASA's space shuttle *Columbia* in 1983. This laptop is the first device that NASA did not build themselves, but bought commercially, for a space shuttle.

THE NONVOLATILE BUBBLE MEMORY HAD NO MOVING PARTS. THE ELECTROLUMINESCENT DISPLAY MADE THE SCREEN VISIBLE EVEN IN SUNLIGHT. THIS MADE IT PERFECT FOR SPACE FLIGHT!

NASA MODIFIED THE GRID COMPASS SLIGHTLY FOR ZERO GRAVITY AND CODE-NAMED IT SPOC (SHUTTLE PORTABLE ON-BOARD COMPUTER).

NASA

MIDI –1983

Disco and synth music were all the rage in the 1970s. But combining sounds from different synthesizers and drum machines was a tedious process. To control the onslaught of new instruments and gizmos, MIDI (Musical Instrument Digital Interface) was created in 1983. This meant that individual musical devices could play in step with each other, sort of like a robotic band reading the same sheet music. Home computers were almost immediately used to compose and manipulate music for electronic instruments. Computers such as the Macintosh were an almost perfect fit for these early GUI-based music-track editors.

The modern DAW (digital audio workstation) is closely related to the early MIDI editors. In 1977, Thomas Stockham's company Soundstream created what first resembled a DAW, which allowed audio to be recorded and edited using a PDP-11 minicomputer. By the late 1980s, computers such as the Atari ST or the Mac—when combined with other peripheral devices—could record and mix studio-quality music. They would replace rooms full of analog tape equipment. Now anyone could turn their bedroom or basement into a recording studio!

INFLUENTIAL PEOPLE
IKUTARO KAKEHASHI
1930-2017
HE WAS BORN IN OSAKA, JAPAN.
AT AGE 16, HE STARTED HIS OWN RADIO REPAIR SHOP. HIS BUSINESS EXPANDED INTO ELECTRIC ORGANS.
"MUSIC IS AS OLD AS THE SHEPHERD WITH HIS PANPIPES, BUT NOW IT ALSO IS AS NEW AS THE SPACE AGE."
Roland
HE FOUNDED ROLAND CORPORATION IN 1972, AND THE ROLAND TR-808 (1980) BECAME THE MOST-USED DRUM MACHINE EVER!
WITH ENGINEER DAVID SMITH, HE PROPOSED THE MIDI STANDARD FOR DIGITAL MUSICAL INSTRUMENTS.
CATMULL WAS THE CO-FOUNDER OF PIXAR, AND HANRAHAN WAS ONE OF THE COMPANY'S FIRST EMPLOYEES.
THEY WON A TURING AWARD FOR THEIR WORK IN 3-D COMPUTER GRAPHICS.
"TO BE A TRULY CREATIVE COMPANY, YOU MUST START THINGS THAT MIGHT FAIL."
EDWIN CATMULL
1945-
AND
PATRICK HANRAHAN
1954-
HE WAS PRESIDENT OF PIXAR FOR 30 YEARS.
HE WAS THE LEAD ARCHITECT OF THE RENDERMAN PROGRAM.
JACK TRAMIEL
1928-2012
TRAMIEL IMMIGRATED TO THE U.S. AFTER SURVIVING THE HOLOCAUST. HE JOINED THE U.S. ARMY, AND REPAIRED TYPEWRITERS.
"WE WILL MAKE COMPUTERS FOR THE MASSES, NOT THE CLASSES."
HE FOUNDED COMMODORE IN 1955.
HE RAN ATARI FROM 1984 TO 1996.
PET
SUSAN KARE • 1954-
SHE IS THE GRAPHIC DESIGNER WHO CREATED THE ICONS FOR THE MACINTOSH.
"GOOD DESIGN'S NOT ABOUT WHAT MEDIUM YOU'RE WORKING IN, IT'S ABOUT THINKING HARD ABOUT WHAT YOU WANT TO DO AND WHAT YOU HAVE TO WORK WITH BEFORE YOU START."
SHE CREATED GRAPHICS FOR MANY TECH COMPANIES, INCLUDING NeXT INC., MICROSOFT, IBM, FACEBOOK, AND PINTEREST.

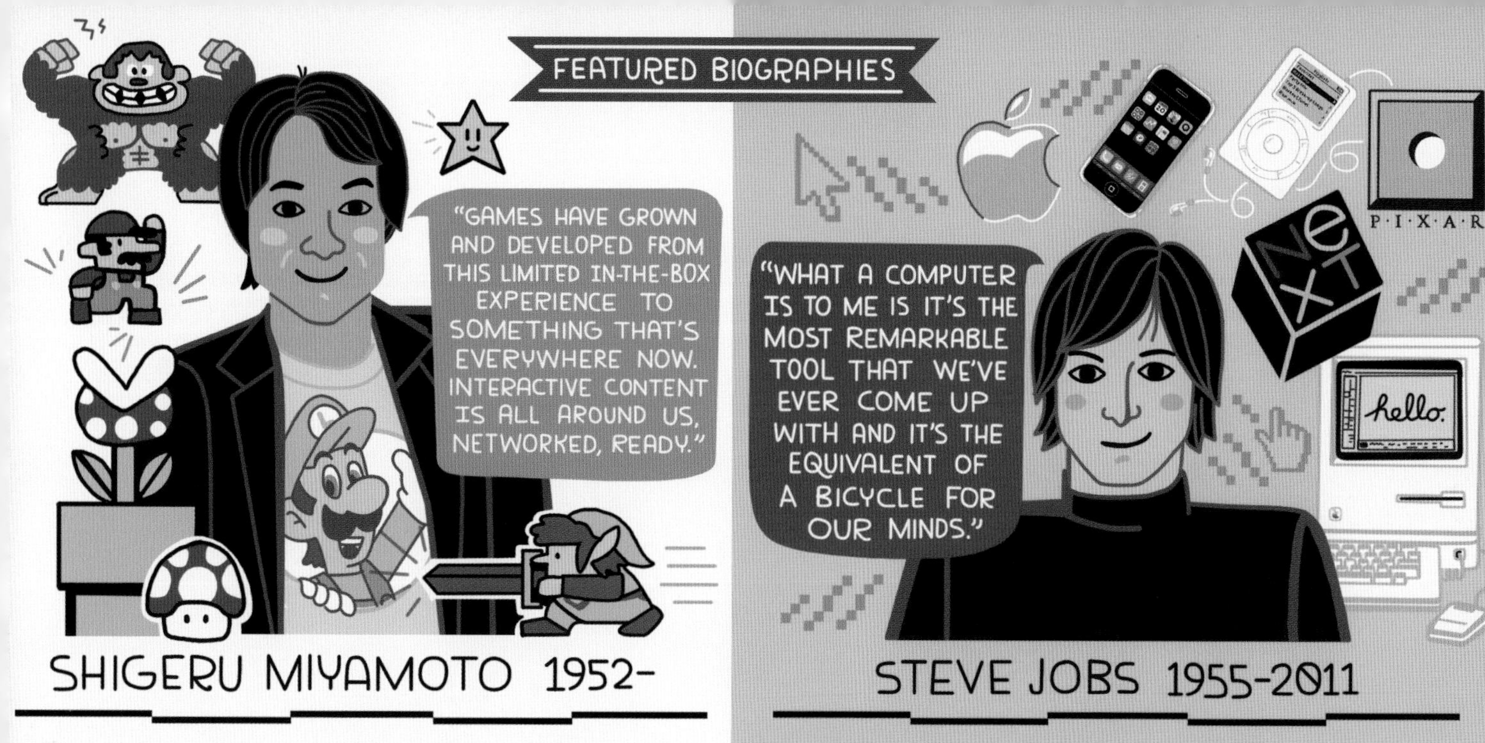

SHIGERU MIYAMOTO 1952-

If you have played popular video games, like *Super Mario Bros.* or *The Legend of Zelda*, you have experienced the artistry of prolific game designer Shigeru Miyamoto. His work was inspired by his adventures as a child, exploring the outdoors at his family's property in rural Japan. As a young man, he had many interests, including playing musical instruments, puppeteering, drawing, and making comic books. He got a job at Nintendo in 1977, and his first success was 1981 arcade game *Donkey Kong*. It was the first video game to have any kind of narrative to play out on a screen, just like a cartoon!

When the 1983 Nintendo Entertainment System arrived, it was as computationally powerful as any home computer of its day. This gave Miyamoto a much larger canvas to re-create his childhood adventure fantasies. In 1985, he created *Super Mario Bros.*, the iconic game where players are led through different side-scrolling worlds, taking video games beyond the static arena-style design that had dominated since the days of *Tennis for Two* (1958). Miyamoto expanded his ideas in *The Legend of Zelda* series, which debuted in 1986 and featured a large, open world for players to explore at their own pace.

Miyamoto's work had the biggest impact of any video-game designer in history. The worlds of game play that he created have since inspired a generation of game designers, writers, user-interface developers, and artists.

STEVE JOBS 1955-2011

Steve Jobs was undoubtedly the most persuasive salesman of his generation and an icon of computer history, even though, according to his business partners, he never learned how to write code. Jobs's greatest strength was to align himself with talented designers, engineers, and artists and to build an environment where they could innovate and create.

Jobs was influenced by European and Japanese minimalist design, incorporating that atheistic into any product line he could. While other computer makers of his era focused on bottom-line sales, Jobs went in a creative direction. He considered the Macintosh a mass-produced work of art akin to a Tiffany lamp, to the point of having all of its designers' signatures embossed inside the case.

During the Mac's development, Jobs' uncompromising style caused friction at Apple, and he left the company in 1985. That same year, he formed NeXT, Inc., and hired many from his former Apple team. A year later, Jobs co-founded Pixar and oversaw its success as an Oscar-winning, industry-leading animation studio.

In 1997, Apple rehired Jobs as its CEO, and his leadership brought the business back from the brink of failure. His biggest triumphs were in taking other companies' murky ideas or creative flops and refining them. He oversaw the development of the iPod (2001), the iPhone (2007), and the iPad (2010) and his user-centric design style helped popularize personal computers and smart devices.

AIM (AOL INSTANT MESSENGER), 1997
iPOD, 2001
FIRST FULLY CGI FEATURE FILM: TOY STORY, 1995
KODAK DCS 420 EARLY DIGITAL CAMERA, 1994
GEOCITIES WEBSITES STARTED IN 1994
PALM PILOT, 1997
CatNY87(8:01AM): Hi! How are you
CALiCAT87(8:01AM): doing good! u?
CatNY87(8:01AM): My CAT ATE my pizza :(
CALiCAT87(8:01AM): So did mine! :)
Playlists
Favorites
Road Trip
Party Mix
Top 5 Break-up Songs
Workout Tunes
Rap mix
Nikon
N90s
Kodak
Welcome to Wienerville
PalmPilot

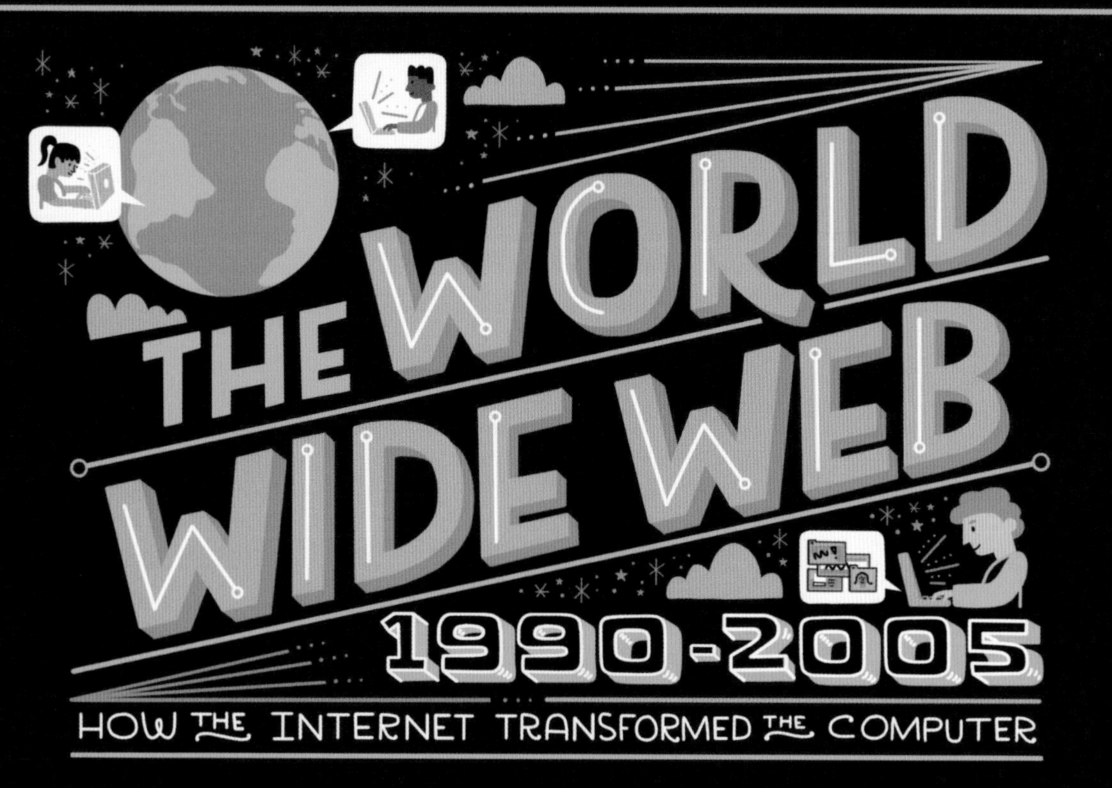

Before the 1990s, computers were like toolboxes. They sat in the office or the "computer room" and were booted up to do specific tasks. For example, a home computer was turned on when Mom needed to use spreadsheets to do taxes or when the kids wanted to play a computer game, like *The Oregon Trail*. When the task was complete, the computer was turned off, and the "toolbox" was put away. In 1990, only 15 percent of U.S. homes had a computer, but that number changed with the creation of the World Wide Web, which made the internet user-friendly. For the first time, everyday people could use their computers to "surf" the Web!

The internet transformed the computer into a multimedia device, allowing people to log on to do (previously offline) tasks, like reading the news, watching videos, or connecting with friends. The internet grew into a vast ever-updating encyclopedia, a global marketplace, and a communication hub—all rolled into one. Internet access became a necessity and was the reason that many homes bought their first computer. By 2000, more than half of U.S. households had a computer.

The 1990s have been called "the Wild West of the internet"—a time of online get-rich-quick schemes, pop-up ads, and brand-new business models. In the early 2000s, "content creators" and "users" began to blend on a grand scale, creating new types of online spaces, like social networking sites. The 1990s and early 2000s saw booms and busts, and the experimentation, successes, and failures of this era shaped the world that we live in today.

TIMELINE

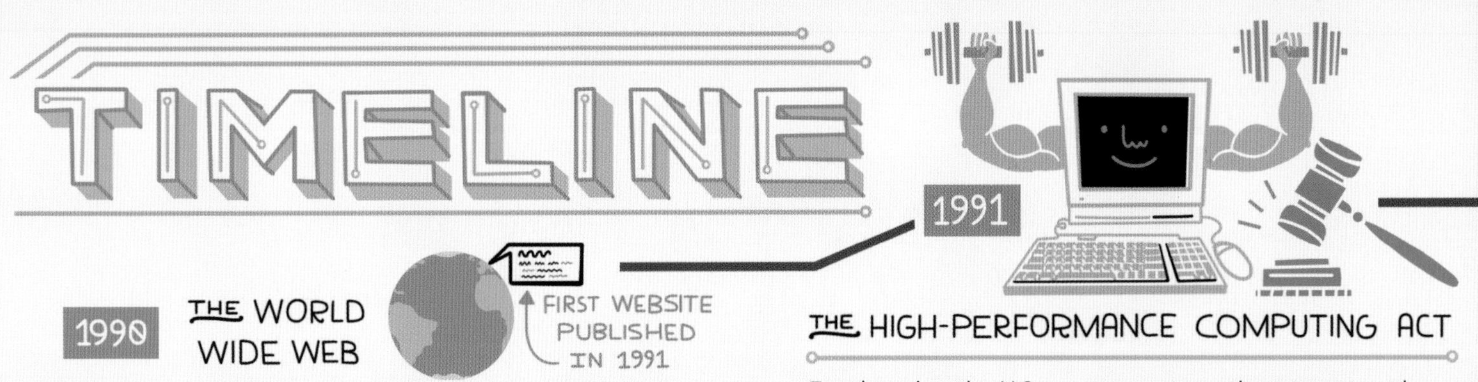

1990 THE WORLD WIDE WEB

The internet was not easy to use until computer scientist Tim Berners-Lee invented "the Web," with help from informatics engineer Robert Cailliau. The Web is an application that runs on top of the internet and uses hypertext to link together different published "web documents."

1991 THE HIGH-PERFORMANCE COMPUTING ACT

For decades, the U.S. government ran the internet, with no commercial activity allowed. In 1991, the High-Performance Computing Act provided $600 million in funding to further develop supercomputing. Commercial traffic was officially allowed on the internet for the first time, leading to the eventual privatization of the internet.

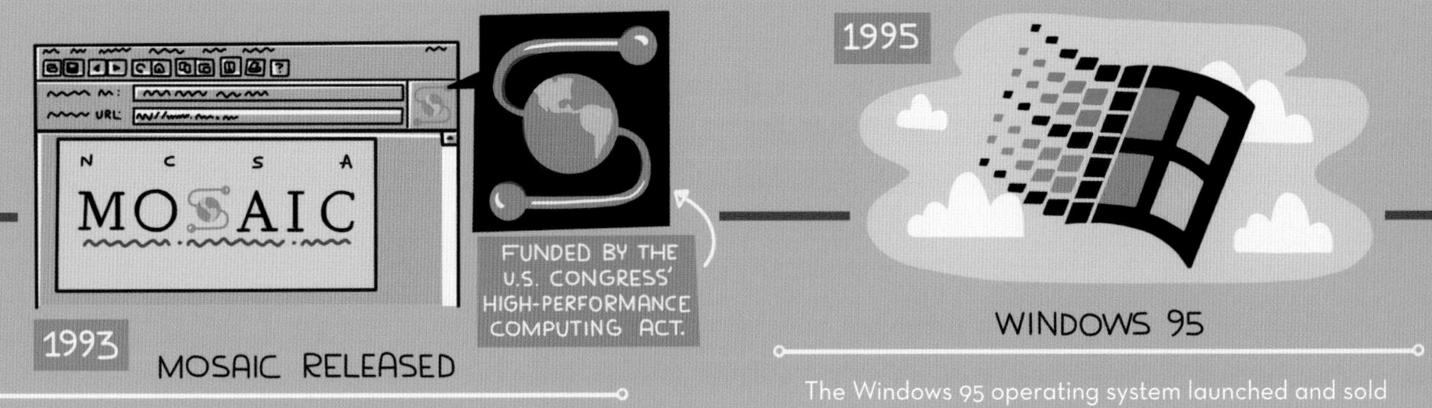

1993 MOSAIC RELEASED

A browser is a software application that allows a person to access the Web by retrieving the requested webpage from a server. Mosaic was the first professional graphical browser that was widely distributed.

1995 WINDOWS 95

The Windows 95 operating system launched and sold more than one million copies in the first four days. Most people were introduced to the internet through Microsoft's Internet Explorer browser that was preloaded onto the Windows 95 desktop.

2000 THE DOT-COM CRASH

In the late 1990s, a lot of money was made investing in dot-com companies. The problem was that most of these companies were extremely overvalued. Many dot-coms went public before having any way to make a profit. They had flashy ad campaigns and massive speculative worth—but no real business. In 2000, the dot-com bubble "popped" and the stock market crashed.

1991 THE LINUX KERNEL

A free operating system called Linux, based on UNIX, was released to Usenet newsgroups on the internet. Immediately, thousands of volunteers began improving it, and, in 1992, it became a popular open-source operating system.

1992 THE JPEG STANDARD

Members of France's Minitel network got together at the Joint Photographic Experts Group (JPEG) to figure out how to compress an image to look good on the internet. In 1992, the JPEG standard was introduced and became one of the most-used file formats!

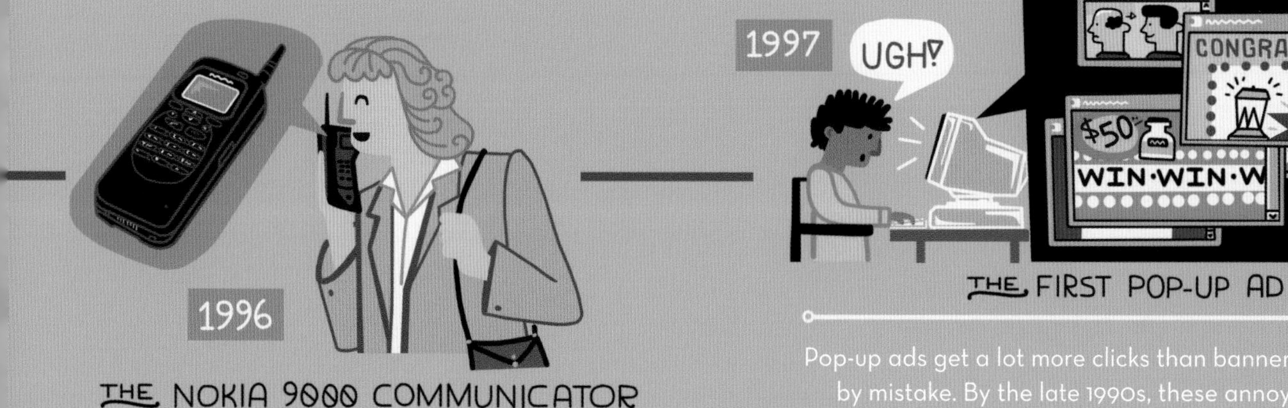

1996 THE NOKIA 9000 COMMUNICATOR

Launched in Finland, this is considered the first cellphone that had internet access.

1997 THE FIRST POP-UP AD

Pop-up ads get a lot more clicks than banner ads, probably by mistake. By the late 1990s, these annoyances were clogging everyone's browsers, leading to a programming arms race between advertisers and scammers and the browser developers trying to block them.

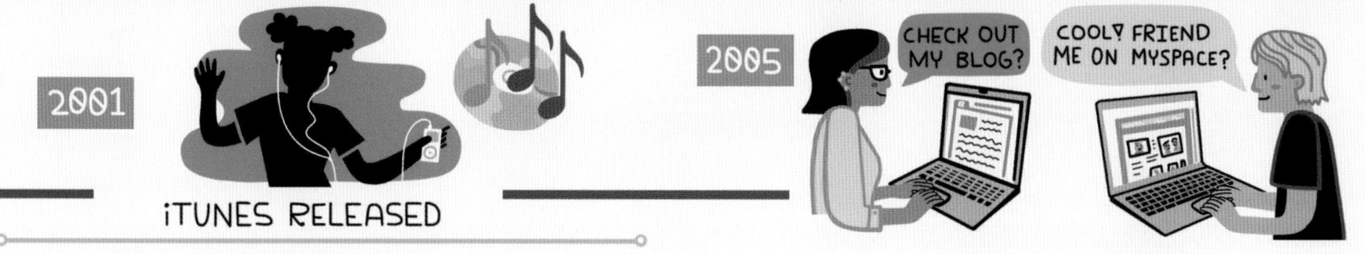

2001 iTUNES RELEASED

In the 1990s, music could be compressed into digital files called MP3s and downloaded online. This completely disrupted the music industry. People no longer needed cassette tapes or CDs. Apple saw an opportunity to make a deal with the record industry and sell music digitally. With iTunes, people could purchase individual songs and listen to them on an Apple iPod.

2005 WEB 2.0 ARRIVED

Many historians call the period of 1991 through 2004 "Web 1.0." By 2005, the Web was no longer primarily made up of static websites, but instead had more user-created content that was constantly updated. This change was called Web 2.0.

STORIES FROM HISTORY

UPLOADED IN 1992, ONE OF THE FIRST KNOWN IMAGES ON THE WEB WAS A PHOTO OF LES HORRIBLES CERNETTES, AN ALL-GIRL DOO-WOP BAND THAT ALSO WORKED AT CERN.

AOL GAVE OUT FREE CD-ROMs BY MAIL AND EVERYWHERE FROM BASEBALL GAMES TO SUPERMARKETS. THEY EVEN TRIED TO FLASH-FREEZE THEM AND BUNDLE THEM WITH STEAKS!

The terms *internet* and *Web* are often used interchangeably, but they are two different things! The internet is the physical network of connected computers, with standards of how data moves across that network. The Web is an application that runs on top of the internet and is a collection of pages, documents, and resources that are all linked together in a "web" of hyperlinks and addresses.

The internet was started in 1969 as U.S. military project ARPANET, which ran "the backbone of the internet" until the late 1980s, when it was taken over by NSFNET. Although other smaller networks popped up around the world, the main internet was heavily restricted by the government agencies that ran it. NSFNET was intended strictly for science and academic research. It was free, and such activity as advertising, commerce, and charging for internet access was not allowed. Other than email, the early internet was pretty much an exclusive club for academics, and it was not user-friendly. Imagine having to scroll through a very long directory of files with huge and incomprehensible names each time you wanted to read a different article. The internet's true potential had not been realized.

WALLED GARDEN

IN THE EARLY 1990s, FIRST-TIME INTERNET USERS FELT OVERWHELMED. "WALLED GARDENS" WERE CLOSED NETWORKS, LIKE AOL (AMERICA ONLINE), THAT PROVIDED LIMITED CONTENT LIKE EMAIL, SPORTS, NEWS, AND GAMES.

THE WORLD WIDE WEB

Tim Berners-Lee created the Web while he was working in Switzerland as a software engineer at CERN (the European Organization for Nuclear Research). Scientists at CERN used NSFNET to send emails and work on particle physics research. While walking around CERN, Berners-Lee was inspired by how fast news traveled through the hallways. In those centralized places, people overheard conversations, read flyers on bulletin boards, and bumped into colleagues, all leading them to share ideas and collaborate. Berners-Lee wanted the internet work like that! In 1989, he formed his proposal for the World Wide Web.

The Web connected related "web documents" through clickable links that were embedded into key words, called hypertext. A mouse click could easily navigate from one document to another. Web documents—later called *webpages* or *websites*—were created using HTML (HyperText Markup Language), a programming language created by Berners-Lee. With the help of Robert Cailliau, the World Wide Web was born in 1990 and released to the public for free use in 1991.

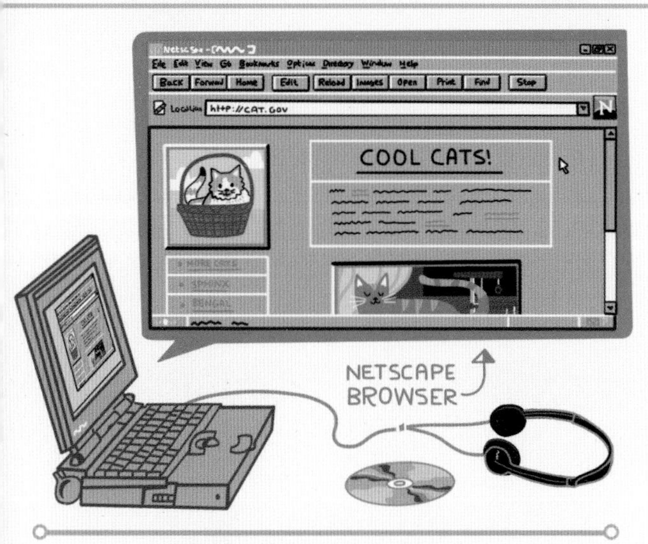

THE BROWSER

A browser retrieves and displays a website. In 1990, Tim Berners-Lee released a browser/page editor named Nexus. This browser was limited to text-heavy webpages, which looked more like a text document than a modern website. In 1993, CERN and Berners-Lee released the source code on the internet to crowdsource improvements to the Web.

That same year, Marc Andreessen and Eric Bina at the University of Illinois co-authored Mosaic, one of the first modern browsers. It was easily installed on a variety of computers and allowed images and colors to be effortlessly incorporated directly into webpages, which now looked more like magazine pages than academic papers. Within a year of its release, tens of thousands of new websites were published.

BROWSER WARS

Marc Andreessen was twenty-two when he led the Mosaic project. After graduating from the University of Illinois in 1993, he started Netscape with entrepreneur Jim Clark. Andreessen's goal was to improve upon his original browser and build a "Mosaic killer." This browser, called Netscape Navigator, was released in October 1994, and it was downloaded more than six million times. The next year, Netscape went public and its stock skyrocketed—the small company was suddenly worth billions. The Web became the new frontier to get rich quick.

This caught the attention of Bill Gates and Microsoft. Gates was worried that Netscape Navigator could eventually replace Windows. It was clear that the future of computers was access to websites and browser-based software. In the mid-1990s, Microsoft nearly had a monopoly in OS market share. Windows 95 was the most anticipated software launch in history, and it came bundled with Microsoft's browser, Internet Explorer.

Netscape and Microsoft competed by improving their browsers, which included adding media features that allowed people to view animations and stream audio. Microsoft even went so far as to force Windows to use only Explorer, by blocking the installation of Navigator. This led to a U.S. government investigation of Microsoft, as well as an antitrust lawsuit. Even though Microsoft was forced to resume allowing the use of other browsers, the damage had been done; Netscape never recovered a significant market share.

Internet browsers are just one example of the tendency for monopolies to form in the tech industry. The most creative and prolific times in computer history happen when competition, different ideas, and active collaborations can thrive. Those fruitful times are then usually followed by creative stagnation when monopolies go unchecked.

THE WINDOWS 95 LAUNCH CAMPAIGN COST $200 MILLION, WITH $3 MILLION SPENT ON THE LICENSING TO USE THE ROLLING STONES' SONG "START ME UP."

PHOTOSHOP, AN IMAGE EDITOR, DEBUTED IN 1990.

THIS ENTIRE BOOK WAS ILLUSTRATED IN PHOTOSHOP!

IN 1995, NSFNET WAS DECOMMISSIONED AND "THE BACKBONE" OF THE INTERNET WAS COMPLETELY HANDED OFF TO PRIVATE COMPANIES.

THE SEARCH ENGINE

The search engine was a big step forward in making the Web user-friendly. By the mid-1990s, the number of websites had skyrocketed. Sure, it was easy to remember or "bookmark" the URLs of your favorite websites, but what if you wanted to find something new? How could you find a website without having the address?

The first modern search engine, JumpStation, was created in 1993 by Jonathan Fletcher at the University of Stirling in Scotland. It had an automated program called a "webcrawler" that indexed the content of every webpage on the internet. That information was then stored on a server. Want to search for information about scuba diving? The content on the server would be sorted using the key words "scuba diving" and a list of relevant websites would be provided.

Most early search engines worked the same way. Websites were not sorted by *quality* of information but rather by the *quantity* of how many times a key word appeared. To increase traffic to their websites, people began cheating the system by adding key words to their sites over and over, sometimes cleverly hidden behind images. Imagine a webpage that was spammed with the word *scuba* a million times being the top result in a search. Not helpful at all!

Two Stanford University graduate students named Larry Page and Sergey Brin solved this problem. Inspired by how an academic paper's quality was determined by how many other published papers used it as a source, they created an algorithm that measured the number of times a website was linked to other websites. This is called backlinking. The more backlinks a website had, the higher it would rank in a search. Originally their search engine started in 1996 as a research project called BackRub. In 1998, Page and Brin began a company and search engine called Google, named after the word *googol*, a "gigantic number."

With such a huge audience, search-engine companies have made a fortune by selling ad space and offering "sponsored search results." Search engines, like Google, also have access to massive amounts of user data, which is a priceless commodity to both advertisers and governments. Google's algorithms worked so well that it survived the 2000 dot-com market crash, and its successful IPO in 2004 signaled to investors that internet companies were still moneymakers. Google has grown into one of the most powerful corporations on the planet.

COMPUTERS CREATED PHOTOREALISTIC CGI IN MOVIES: *TERMINATOR 2* (1991)

JURASSIC PARK (1993)

THE MATRIX (1999)

STARTED IN 1996, ROBOCUP IS AN INTERNATIONAL ROBOTICS AND AI SOCCER COMPETITION.

THE FIRST BANNER AD

HOW TO MAKE MONEY ONLINE WHEN EVERYTHING IS EXPECTED TO BE FREE? ADS!

UNLIKE PRINT OR TV ADS, MARKETERS KNOW HOW MANY PEOPLE CLICKED ON AN ONLINE AD AND HOW THOSE CLICKS RESULTED IN SALES. THIS DATA WAS VERY VALUABLE, AND WAS JUST THE START.

SELLING AD SPACE AND DATA COLLECTED ON USERS IS HOW MANY ONLINE COMPANIES CONTINUE TO MAKE MONEY.

EARLY SOCIAL MEDIA

In the early days of the Web, there was a divide between content creators and consumers, with a small amount of people building websites for millions of visitors. To have your own website, you needed either to be tech-savvy enough to code in HTML or have the money to pay a developer. This began to change when the GeoCities platform launched in 1994. With GeoCities, people could easily create their own webpages from a basic set of online tools and little coding. At the peak of its popularity in 1999, GeoCities had more than thirty-eight million pages, each categorized in "neighborhoods" by topic. This started to blur the lines between content creator and consumer.

GeoCities was a precursor to the modern social network. In the 2000s, early social media platforms such as Friendster (2002) and Myspace (2003) went online. Multiple times a day, users would return to these sites to post messages to each other and update their own semipublic webpages. In 2004, Harvard University student Mark Zuckerberg created Facebook to connect students at his university. Over the next few years, Facebook expanded to other universities and then rapidly to the entire world, until it became a massive media conglomerate.

IN COLLEGE, MARK ZUCKERBERG STARTED A SITE CALLED FACEMASH TO RATE THE LOOKS OF FRESHMEN.

HARVARD SHUT IT DOWN BECAUSE IT WAS SEXIST.

ON MYSPACE, CEO TOM ANDERSON WAS EVERYONE'S FIRST "FRIEND."

"ALL RIGHT, SO HERE WE ARE IN FRONT OF THE ELEPHANTS..."

THE FIRST YOUTUBE VIDEO, "ME AT THE ZOO," WAS UPLOADED IN 2005.

IMPACT OF THE ERA

The internet grew into a vast and ever-updating encyclopedia, a global marketplace, and a communication hub—all rolled into one. By the end of 2005, the Web had matured, and "users" and "content creators" were one and the same. Web 2.0 had arrived, setting the stage for the next decade, where the internet would become an extension of personal identity.

BUILD A WEBSITE!

JOIN A CHAT GROUP!

FOR THE FIRST TIME, PEOPLE COULD EASILY FIND COMMUNITIES FOR THEIR OBSCURE OBSESSIONS, AND EVEN BUILD BUSINESSES ABOUT THEM!

START A LIVE STREAM!

FINALLY SOMEONE WHO ALSO LOVES BATS!

IMPORTANT INVENTIONS

BROADBAND FOR THE INTERNET–1996

Before broadband, the internet was accessed through telephone lines via a modem that translated analog signals into digital signals. This was called "dial-up" internet access. It was a system built on a communication infrastructure designed for telephone calls in the 1950s, not for massive amounts of digital data. This meant the internet was very slow, especially when transmitting photos and other large files.

Broadband refers to high-speed internet that is faster than dial-up. Broadband uses various ways to transfer data, including cable, optical fiber, and satellite. In 1996, a cable-modem service in Canada introduced broadband to North America, but it didn't become widespread until the early 2000s. By 2010, 65 percent of U.S. households used broadband to get online. Most of the internet access in rural areas, however, continued to use a much slower dial-up connection. With internet speed comes power! The quick transmission of large files means a more useful internet!

WIKIPEDIA–2001

Jimmy Wales and Larry Sanger created Wikipedia, a free online encyclopedia, in 2001. When it first started, it was called Nupedia and contained articles written by experts that were formally reviewed for accuracy. While working on Nupedia, Wales had the idea to create a publicly written encyclopedia that could contain information on any subject and be editable by anyone in the world. This method of obtaining and editing information is called crowdsourcing. Wikipedia grew quickly—by 2007 it had passed the two million article mark, making it the largest encyclopedia to ever exist. It continues to grow and remains one of the most-visited sites on the internet.

Wikipedia offers information about an enormous range of topics and is a great place to get a general understanding of an issue. However, online libraries such as Wikipedia are created by the masses, and with that can come mistakes, bias, and misinformation. Today's Wikipedia has rules and administrators who flag inaccuracies. This helps keep the articles factual, although not always perfect. When doing research online, it's a good idea to verify information using multiple trustworthy sources and to consult well-researched publications.

POCKET DEVICES, PERSONAL ASSISTANTS, AND MP3 PLAYERS

In the 1980s, cellphones were bulky, expensive, and often built into cars, so they were mainly used only by the wealthy. By the late 1990s, cellphones had become small and inexpensive enough for the average customer. Tech companies wanted to take it a step further and create pocket devices that did more! Inspired by the hand-size communication gizmos in detective comics of the 1940s and *Star Trek*, designers tried (with mixed success) throughout the 1990s and early 2000s to create an all-in-one communication device.

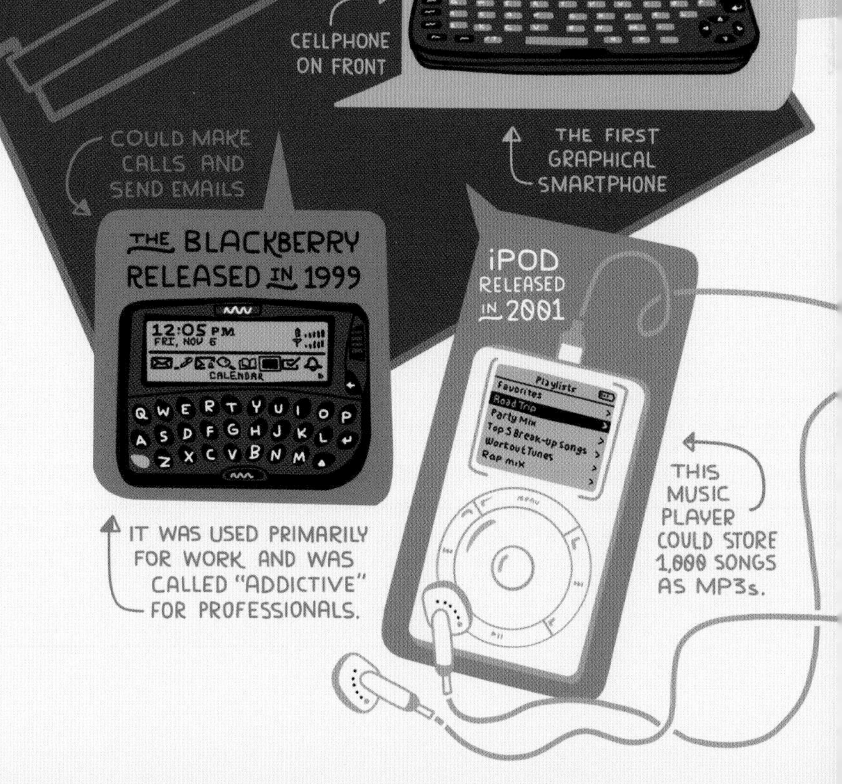

INFLUENTIAL PEOPLE
111
001011
1000
001011
1111000
TOGETHER THEY WON A TURING AWARD FOR THEIR WORK IN CRYPTOGRAPHY.
SILVIO MICALI
1954–
AND
SHAFI GOLDWASSER
1959–
THEIR PAPER "PROBABILISTIC ENCRYPTION" WAS CRUCIAL TO THE DEVELOPMENT OF SECURITY ON THE INTERNET.
110
00101100
1000
WITH CHARLES RACKOFF, THEY ARE THE CO-INVENTORS OF ZERO-KNOWLEDGE PROOFS, WHICH ARE KEY TO DESIGNING CRYPTOGRAPHIC PROTOCOLS.
LINUS TORVALDS
1969–
"THAT'S WHAT MAKES LINUX SO GOOD: YOU PUT IN SOMETHING, AND THAT EFFORT MULTIPLIES. IT'S A POSITIVE FEEDBACK CYCLE."
BORN IN HELSINKI, FINLAND.
HE CREATED THE FREE OPERATING SYSTEM LINUX IN 1991.
LINUX IS ONE OF THE MOST POPULAR OPEN-SOURCE OPERATING SYSTEMS. VERSIONS OF LINUX ARE FOUND WORLDWIDE ON SMARTPHONES.
THE QUIRKY AND CHAOTIC ATMOSPHERE OF EBAY SHAPED EARLY WEB CULTURE.
ebay
HE'S A FRENCH-BORN, IRANIAN AMERICAN SOFTWARE ENGINEER.
"WHAT MAKES EBAY SUCCESSFUL—THE REAL VALUE AND THE REAL POWER AT EBAY—IS THE COMMUNITY. IT'S THE BUYERS AND SELLERS COMING TOGETHER AND FORMING A MARKETPLACE."
PIERRE OMIDYAR · 1967–
IN 1995, OMIDYAR LAUNCHED AUCTIONWEB, WHICH WAS RENAMED EBAY. ALTHOUGH IT STARTED AS A HOBBY, BY 2001, EBAY HAD BECOME ONE OF THE LARGEST E-COMMERCE SITES ON THE WEB.
WITH JEANNETTE WING, SHE DEVELOPED THE LISKOV SUBSTITUTION PRINCIPLE, WHICH IS A PRINCIPLE IN OBJECT-ORIENTED PROGRAMMING.
EARNED A TURING AWARD FOR HER FOUNDATIONAL WORK IN PROGRAMMING LANGUAGE AND SYSTEM DESIGN.
SHE LEADS THE PROGRAMMING METHODOLOGY GROUP AT MIT.
BARBARA LISKOV
1939–

"THE WEB AS I ENVISAGED IT, WE HAVE NOT SEEN IT YET. THE FUTURE IS STILL SO MUCH BIGGER THAN THE PAST."

SIR TIM BERNERS-LEE 1955-

British computer scientist Sir Tim Berners-Lee was born into a tech-savvy family—both of his parents worked on the United Kingdom's first commercial computer, the Ferranti Mark 1. After Berners-Lee graduated from Oxford in 1976, he had several jobs developing software. In 1980, he spent a few months as a software consultant for CERN in Switzerland, where he developed a program, called ENQUIRE, that used hypertext links. Four years later, he returned to CERN and was tasked with developing the lab's computer network. He wanted to figure out a better way for the scientists to share data and ideas; this led to his 1989 proposal for the World Wide Web. "The original idea of the Web was that it should be a collaborative space where you can communicate through sharing information," Berners-Lee said. Completed in 1990, the Web was made public the following year.

Berners-Lee fought to make the Web completely free to use—as he put it, to "let a thousand flowers bloom" and inspire innovation. In 1994, he established the World Wide Web (W3) Consortium, which is an international community that develops Web standards. It was created with the goal to ensure that the Web continues to grow and remains in the public domain. In 2004, Berners-Lee was knighted in recognition of his contributions. Everything from online businesses (both the very small and the Web-tech giants), crowdsourced research, community forums, and personal blogs are all possible because of the open-source Web that Berners-Lee created.

BILL GATES 1955-

In the 1990s, Microsoft had an undisputed monopoly on operating systems, and founder Bill Gates was at the height of his power and celebrity. In 2000, more than 97 percent of computers used Windows OS, and Microsoft Office and Internet Explorer were universal standards.

Gates was born in Seattle, Washington. He and his childhood friend Paul Allen experimented with programming, using a time-sharing terminal at their high school. Together, they built a very simple traffic data-collection computer called Traf-O-Data. This gave them programming experience with Intel microprocessors that few people had at the time. In 1975, Gates followed Allen's lead and dropped out of Harvard to work on adapting BASIC for the Altair 8800, an early home computer. That same year, in Albuquerque, Gates and Allen co-founded Microsoft. Their programming language, Microsoft BASIC, was adapted to almost every microcomputer in the 1970s, giving them an early foothold in a highly competitive industry.

Gates continued to grow the company in the 1980s. Microsoft became the industry standard by collaborating with IBM on their PC. Microsoft's most famous product, the Windows GUI, first launched in 1985. Under Gates's leadership, Microsoft went for the same reach and market dominance that IBM had been known for in the 1960s. For a while, having a personal computer meant using Windows OS almost without exception.

In 2000, Gates and his then wife launched the Bill & Melinda Gates Foundation, and he turned his attention to philanthropy. The foundation primarily focuses on health care, education, and fighting poverty and is the largest private foundation in the world. In 2008, Gates left his role at Microsoft to dedicate himself to the foundation.

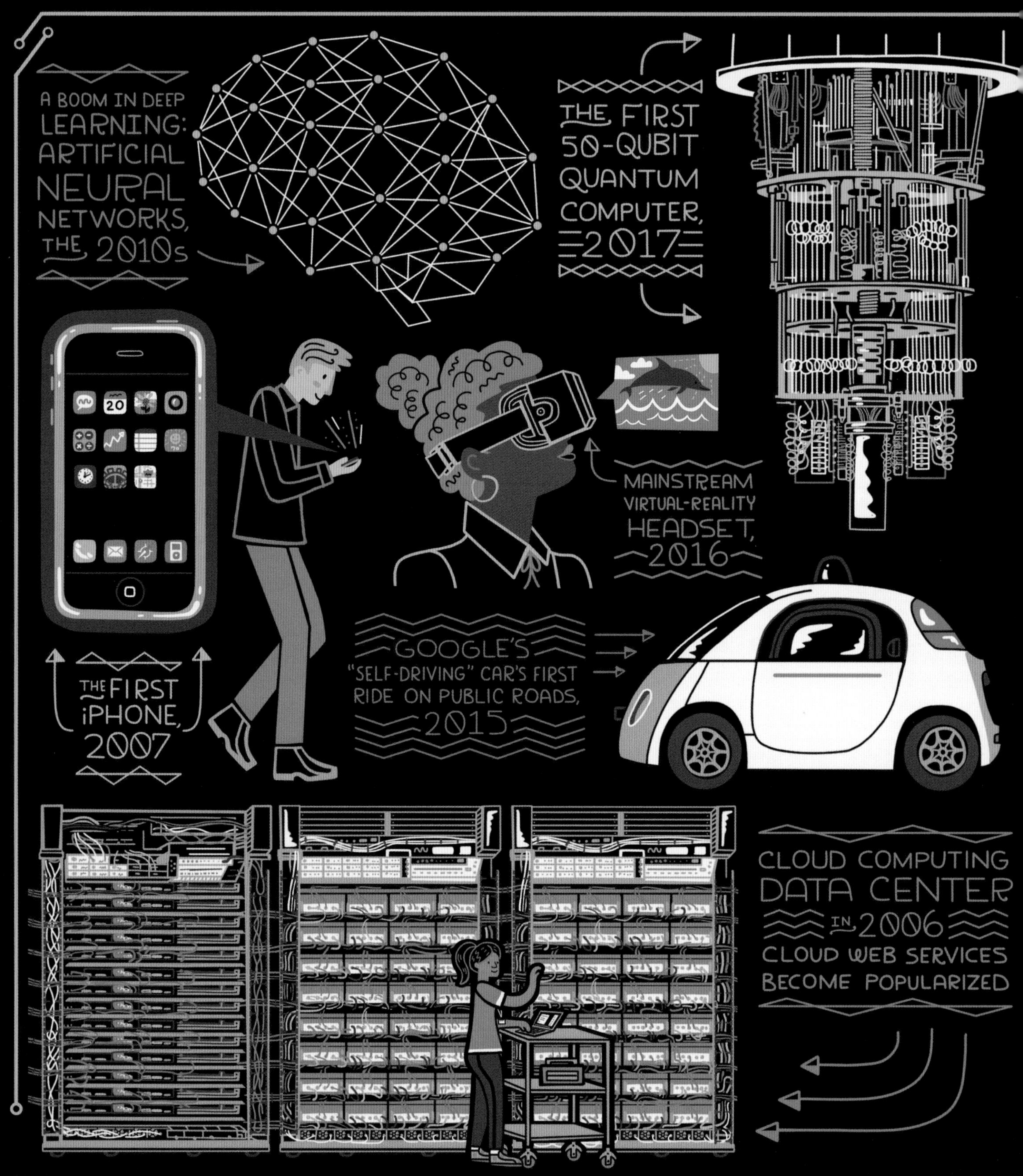
A BOOM IN DEEP LEARNING: ARTIFICIAL NEURAL NETWORKS, THE 2010s
THE FIRST 50-QUBIT QUANTUM COMPUTER, 2017
MAINSTREAM VIRTUAL-REALITY HEADSET, 2016
THE FIRST iPHONE, 2007
GOOGLE'S "SELF-DRIVING" CAR'S FIRST RIDE ON PUBLIC ROADS, 2015
CLOUD COMPUTING DATA CENTER
IN 2006
CLOUD WEB SERVICES BECOME POPULARIZED

PORTABLE COMPUTERS, BIG DATA, AND AI

By 2005, Web 2.0 had arrived! The computer had become a vital tool for communication, work, and play. In just a few years, the smartphone would gain dominance. This all-in-one device combined a cellphone, computer, digital camera, GPS, and more—and would become an indispensable tool for constant internet access. The rise of the smartphone created a huge cultural shift in developed countries, where life became increasingly filtered through interactions powered by the internet.

Wi-Fi and Bluetooth technology, created in the 1990s, had fully matured in the first decade of the 2000s and became increasingly common in consumer technology. Many household appliances—thermostats, refrigerators, security systems, and more—became a "smart" device by connecting to the internet.

During the 2010s, portable smart devices and wireless broadband became commonplace. The internet was no longer a destination accessed by sitting in front of a desktop computer, instead it was omnipresent. Gigantic data centers were used to house the massive amount of information created online. Huge amounts of data, and ever increasingly powerful computers, allowed the advancement of AI with complex neural networks. Computer technology in the next decades would be defined by huge leaps in AI research.

TIMELINE

2006

GOOGLE ADDED TO THE DICTIONARY

The *Oxford English Dictionary* and *Merriam-Webster Dictionary* both added the verb "google." Google continues to be the way a majority of people navigate the internet. Small changes to Google's search algorithm can have big impacts on what people find online.

2012

RASPBERRY PI

The best way to learn about computers is to program one! The Raspberry Pi Foundation built small computer boards for students to modify and program, using languages such as Scratch and Python. By 2013, more than a million Raspberry Pi computers were being used as teaching tools.

2014

GOOGLE NEURAL MACHINE TRANSLATION

With NMT (neural machine translation) software, translators are able to read whole sentences at once to understand the correct word endings, tense, or plurals. It is a large improvement over older translators that would read only one word or phrase at a time.

2015

ALPHABET INC.

Google continued to grow into one of the largest and most powerful tech companies on the planet. In 2015, Google formed the Alphabet Inc. conglomerate and within a year had bought more than two hundred companies. Alphabet includes companies that focus on health care, AI, self-driving cars, internet access, and more.

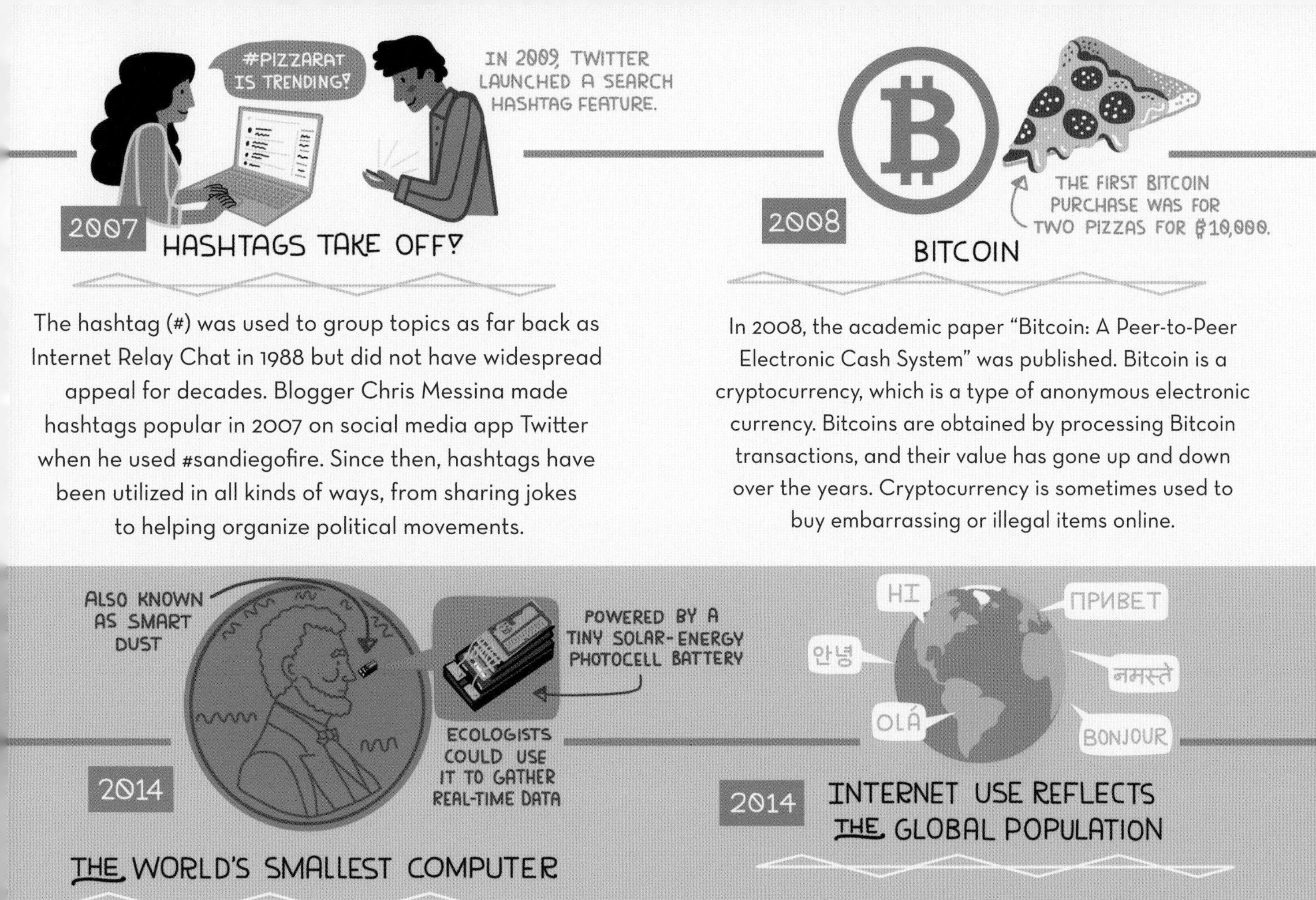

2007 HASHTAGS TAKE OFF!

The hashtag (#) was used to group topics as far back as Internet Relay Chat in 1988 but did not have widespread appeal for decades. Blogger Chris Messina made hashtags popular in 2007 on social media app Twitter when he used #sandiegofire. Since then, hashtags have been utilized in all kinds of ways, from sharing jokes to helping organize political movements.

2008 BITCOIN

In 2008, the academic paper "Bitcoin: A Peer-to-Peer Electronic Cash System" was published. Bitcoin is a cryptocurrency, which is a type of anonymous electronic currency. Bitcoins are obtained by processing Bitcoin transactions, and their value has gone up and down over the years. Cryptocurrency is sometimes used to buy embarrassing or illegal items online.

2014 THE WORLD'S SMALLEST COMPUTER

Computer scientists at the University of Michigan built three computers, called Michigan Micro Motes (a.k.a. M3s), each one the size of a grain of sand. One measured temperature, one measured pressure, and the third could take photos.

2014 INTERNET USE REFLECTS THE GLOBAL POPULATION

By the mid-2010s, global computer use had grown to the point where internet users better reflected the world's population proportionally. In 2014, Chinese citizens became the largest user group on the internet.

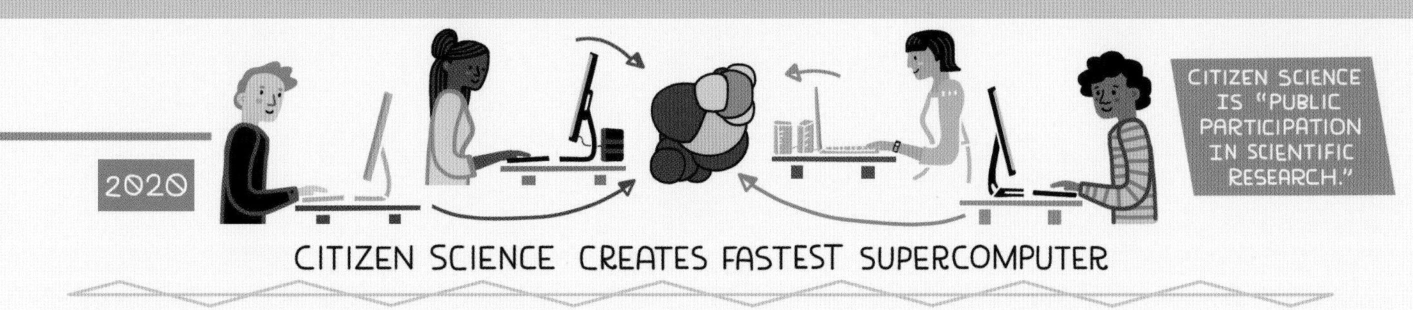

2020 CITIZEN SCIENCE CREATES FASTEST SUPERCOMPUTER

Not every science lab had access to the supercomputers needed to model complicated structures, such as folding proteins of a virus. Stanford University's Folding@home program used the internet to network people's personal computers and share their processing power. In the past, Folding@home has been used to model such viruses as HIV and Ebola. During the COVID-19 pandemic, nearly a million new individuals networked their computers to help fight the disease. During a brief period in the spring of 2020, the Folding@home network became the fastest supercomputer on the planet.

STORIES FROM HISTORY

If you had looked inside a typical backpack in the early 2000s, you might have found many different portable electronics—a Game Boy for video games, an MP3 player for music, a cellphone, or a digital camera. All of these were each created to do a specific task. The next step was obvious—smash all those technologies together to make an all-in-one device that would transform the cellphone into a computer.

IN 2007, THE FIRST TABLET SPECIFICALLY MADE FOR E-BOOKS, THE KINDLE, WAS RELEASED.

SMARTPHONES

IBM's 1994 Simon is considered the very first smartphone. It was limited to only thirty minutes of talk time, was the size of a brick, and couldn't send emails wirelessly, making it a complete flop! Over the years, other proto-smartphones—like the Nokia 9000 Communicator (1996) and the BlackBerry 5810 (2002)—were more successful in the market, but still somewhat niche. These devices had small screens and tiny keyboards made of plastic buttons. Smartphones did not have mass appeal because of limited wireless data and low-computing power, but the iPhone changed all of that.

In 2007, Steve Jobs gave an onstage demo of the iPhone to great fanfare. The crowd oohed and aahed as he showed off all it could do—from watching a clip from TV's *The Office* to answering emails, taking photos, playing songs by Green Day on iTunes, and, of course, making a phone call. All of this was done with just a pinch, tap, or swipe on the glass screen. This pop-culture moment created a sensation—people even camped overnight outside of Apple stores to purchase the iPhone when it launched.

The first iPhone was essentially a small tablet computer combined with a cellphone. With its glass touchscreen and user interface, its function could change depending on the program it was running. This innovation meant there were no tiny buttons, which had limited the utility of earlier smartphones, and it offered the potential for all kinds of software applications. iPhones used a language of gestural movements created by Apple's user-interface team and set the design standard for all smartphones that would follow it.

The iPhone happened when the infrastructure for wireless networks had matured. Unlike the smartphones that had come before, the iPhone had enough bandwidth to view video, quickly browse the Web, and support real-time location tracking.

Around the time that the iPhone was released, other companies, like Android (which was purchased by Google in 2005), had also been working on their own mobile operating systems. The first smartphone to run the Android OS was the HTC Dream (2008). Unlike the iPhone's closed iOS model, the Android OS was an open operating system based on Linux. Although Apple was the first to introduce customers to these new devices, Android would become the market leader. Their open-source OS allowed for more "smart" devices beyond cellphones to run on their software. In 2013, Android-based smartphones sold more units than all other smartphones and personal computers combined.

IN 2012, IT WAS PROVED THAT COMPUTER DATA COULD BE STORED AS GENETICALLY ENGINEERED STRANDS OF DNA.

DNA IS MADE UP OF FOUR CHEMICAL BASES KNOWN AS A, C, G, AND T, WHICH CAN REPRESENT 1s AND 0s.

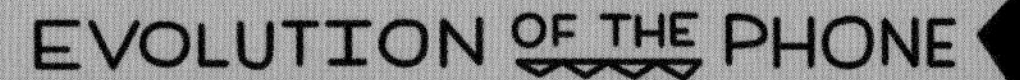

EVOLUTION OF THE PHONE

IN THE 2010s, VIDEO-GAME GRAPHICS AND ONLINE PLAY BECAME A HUGE DRIVER OF COMPUTER INNOVATION.

APPLICATIONS

What truly made smartphones all-in-one were the applications. At first, Apple allowed only certain developers to create "apps" for the iPhone. This didn't seem fair to many people, who saw no reason why such a powerful computer should be reined back. People began hacking (called jailbreaking) into their iPhones to install their own programs. In reaction to this, Apple opened the App Store, where third-party software companies could sell programs with Apple's approval. Throughout the 2010s, start-up tech companies created applications far beyond what iPhone designers had imagined. Apps have since disrupted and changed markets such as transportation, shipping, and health care. With little expense and nearly nonexistent regulations, app companies defined the 2010s as a profit-chasing heyday, similar to the dot-com boom of the 1990s.

IN 2010, THE U.S. AIR FORCE CREATED A POWERFUL SUPERCOMPUTER, BY CONNECTING TOGETHER 1,760 PLAYSTATIONS.

IT WAS NICKNAMED THE "CONDOR CLUSTER."

SMARTPHONES TAKE OVER THE WORLD

A desktop or laptop computer is still required to do serious work, like professional coding, long-form media creation, using business software, and server hosting. Despite this, by 2015, the most common computing device was not the personal computer—it was the smartphone. In 2019, Pew Research estimated that more than half of the five billion mobile users worldwide had smartphones. By 2021, 85 percent of adults in the United States owned a smartphone. At the same time, 15 percent of American adults were dependent on their smartphones for online access and owned no other computer. "Smartphone dependent" users tend to be younger or have lower incomes. While pocket devices are less functional (by design) than a personal computer, they are generally less expensive and more convenient.

The smartphone began a trend of deconstructing the computer—creating consumer electronics with the potential for a lot of computing functionality but limited by design and marketed to do specialized or passive tasks. For example, many people find it easy to read a book on a smartphone, but not many books have been written on one.

When it comes to technology, progress isn't always linear. Modern tablets offer much more computing power than personal computers had in the early 2000s; but in some critical ways, they offer less functionality and control over storage and software. They still have not achieved many of the goals set by the engineers who originally conceptualized tablet computers in the 1970s (for example, Alan Kay's Dynabook). It is important not just to see the past as a precursor to our present but as a rich source of ideas that have not yet been realized.

INTERNET EVERYWHERE

DING

DING

DING

DING

IN THE PAST, THE INTERNET WAS A "PLACE" THAT A PERSON ACCESSED BY LOGGING ON TO A COMPUTER. NOW, PEOPLE ARE CONSTANTLY ONLINE, AND IT'S MORE DIFFICULT TO FIND A PLACE TO "UNPLUG."

IT FEELS GOOD TO UNPLUG!

CLOUD COMPUTING

Despite what its name suggests, cloud computing is not a magical floating mist of data in the air. In reality, "the cloud" refers to rows and rows of computer servers that can store and process massive amounts of data. Cloud service has its roots as far back as 1960s timesharing, when people used terminals to access remote, powerful mainframe computers. Modern servers are kept cooled in gigantic warehouses called data centers, which are reminiscent of old-school mainframes but with exponentially more power. By connecting to these servers via the internet, people can rent storage space and processing power.

It was necessary for large companies such as Amazon and Google to invest in building server infrastructure that could handle data on a global scale. By 2006, cloud computing itself became a big business. Many companies were experiencing storage shortages, and cloud computing allowed them to rent processing time and storage on remote servers owned by the tech giants. As more people were using smartphones, it became common to keep all their personal data remotely on the cloud. This shift away from desktops to portable devices connected to a more powerful server is often described as the Post-PC era.

SOCIAL MEDIA AND ALGORITHMS

Social media platforms evolved throughout the early 2000s to become a one-stop shop for news, community engagement, and entertainment. In some respects, modern social media platforms, like Facebook, are similar to the curated "walled gardens" of the 1990s, like AOL.

Social media networks make money by selling user data to advertisers, researchers, political parties—or anyone!—and by selling ad space and getting eyes on sponsored content.

Such apps all use AI and algorithms based on personal data to present content to users that entices, entertains, or shocks them into staying on a site longer. From the perspective of selling space for advertisements, this works great. However, when people use social media to get the majority of their news, it can be harmful. Poorly written algorithms can lead people to form beliefs based on entertainment and sensationalism rather than on legitimate, fact-checked news sources.

IN 2013, ADOBE STOPPED SELLING PHYSICAL COPIES OF SOFTWARE, MAKING IT ONLY AVAILABLE FOR RENT ONLINE.

THIS STARTED A TREND OF SOFTWARE SUBSCRIPTION SERVICES.

IN 2010, THE FIRST PHOTO ON INSTAGRAM WAS TAKEN BY CO-FOUNDER KEVIN SYSTROM.

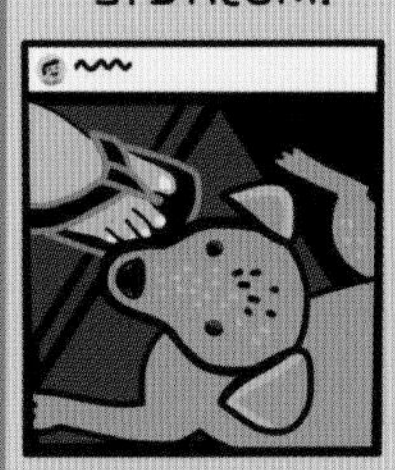

IT WAS OF A DOG AT A TACO STAND.

BIG DATA AND PRIVACY

IN 2012, GOOGLE X FAMOUSLY CREATED A MASSIVE NEURAL NETWORK THAT LEARNED TO DETECT CAT VIDEOS ON YOUTUBE.

THE AI HAD 1 BILLION PARAMETERS.

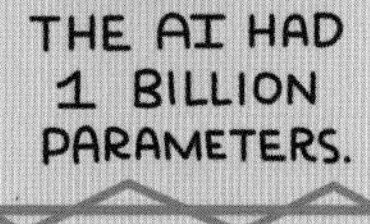

IN 2016, AN AI WAS CREATED TO "PAINT" LIKE BAROQUE ARTIST REMBRANDT USING A 3-D PRINTER.

"Big data" refers to tech companies' ability to process and store billions of users' personal data. The type of data being sold can include photos posted to social media, the amount of time a user spends watching a video, recent online purchases, search history, or even the path a person walks to work every day. Tons of stuff! All of this data is collected through smart devices and by tracking online interactions. It is a massive, exponentially growing amount of information being stored at a rate never before seen in human history.

In addition to being used to create super-specific, targeted advertisements, personal data is also sold and used in scientific and social research, like the development of AI. The ability to store and track user data has allowed for mass online surveillance and has led to many debates about privacy rights.

THE AI BOOM

Is it possible for a machine to learn like a human? This idea has inspired an entire branch of computer science called AI (artificial intelligence). AI is all about getting machines to mimic human behavior or thought. Scientists work to achieve this goal through machine learning, which refers to the different kinds of algorithms and statistical models used to train machines.

Normally, a computer executes a program that gives it step-by-step instructions. Instead, AI and machine learning are all about "teaching" a computer to solve a problem without explicit instructions. A computer "learns" based on an algorithm and by sifting through "training data." Once the computer has processed enough training data, it creates a mathematical model to work on its own.

Interest in AI had many ups and downs, with several periods that historians refer to as an "AI winter," when the field was unpopular and funding for research was cut. By the early 1990s, microprocessors had become fast enough to reinvigorate interest. In the 2010s, a wealth of big data collected from the internet and major developments in machine learning allowed for an explosion in progress. Leaps were made in computer vision, speech recognition, and machine translation.

Most people interact with AI daily without giving it a second thought. AI is at work during a Google image search. It is used on smartphones to autofill text messages and respond to spoken requests. AI can translate languages instantly. We also depend on AI for tasks that would be impossible without a computer's help, such as modeling galaxies and subatomic particles. These are only a few examples of what AI can do!

Although AI mimics how humans learn and problem-solve, they are far from human. In science fiction, AI is often portrayed as able to do any task and hold meaningful conversations, but this type of AI (called general AI) is still many years away.

ARTIFICIAL NEURAL NETWORKS AND DEEP LEARNING

THIS SUBSET OF MACHINE LEARNING IS A POWERFUL TOOL USED IN AI.

ANNs (ARTIFICIAL NEURAL NETWORKS) ARE LOOSELY BASED ON HOW NEURONS TRANSMIT INFORMATION IN THE HUMAN BRAIN. *DEEP LEARNING* REFERS TO MANY HIDDEN LAYERS WITHIN A NEURAL NETWORK.

ANNs WERE ORIGINALLY CONCEIVED IN THE 1940s, BUT WERE CONSIDERED FAR-FETCHED UNTIL THE EARLY 2000s, WHEN PROCESSING POWER AND BIG DATA ALLOWED FOR ANN MODELS TO SCALE UP.

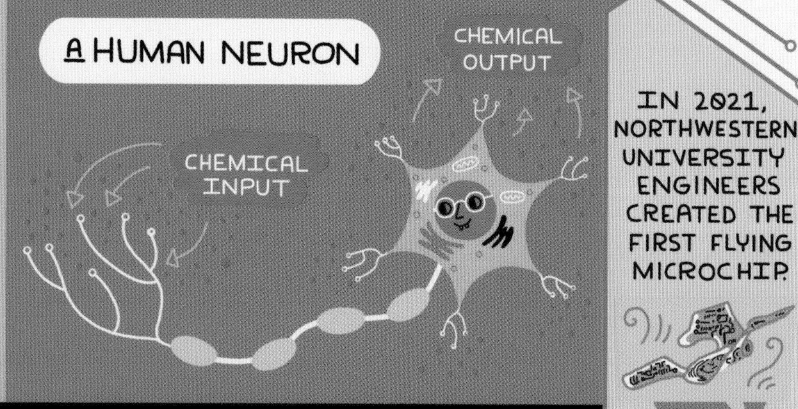

IN 2021, NORTHWESTERN UNIVERSITY ENGINEERS CREATED THE FIRST FLYING MICROCHIP.

THE MICROFLYER IS THE SIZE OF A GRAIN OF SAND AND FLOATS ON THE WIND LIKE A SEED.

A NEURAL NETWORK DIAGRAM

INPUT DATA

NEURON

CONNECTIONS ARE CALLED EDGES.

OUTPUT

OUTPUT LAYER

HIDDEN LAYERS

LAYERS TYPICALLY INFLUENCE ONE ANOTHER BY WEIGHTED CONNECTION.

IMPACT OF THE ERA

Smartphones and constant internet access have allowed for instant communication on a global scale. Far-away families and friends can keep in touch easily and people can stay connected while traveling or working remotely. Online business has become the standard. Thanks to social media, it is common for people to share their lives and work with online audiences. The modern world is being defined by this constant connectedness—of people as well as our household items. The digital revolution is here to stay, bringing with it both great opportunities and challenges.

IN 2021, FEDEX USED AUTONOMOUS TRUCKS FOR DELIVERIES. IT WAS AN "INDUSTRY FIRST."

THESE TRUCKS STILL REQUIRED SAFETY DRIVERS.

IMPORTANT INVENTIONS

VIRTUAL ASSISTANTS GO MAINSTREAM—2011

In 1952, Bell Labs created the first speech-recognition system: a six-foot-tall machine named Audrey (AUtomatic Digit REcognizer) that could recognize the spoken digits of 0 through 9. Computer scientists have since come a long way in voice-recognition technology. In 2011, Apple included the first modern voice-activated virtual assistant, called Siri, on the iPhone OS. At first, Siri could only be used for simple tasks, like sending text messages, checking the weather, or setting an alarm. Within just a few years, Siri was programmed to answer any question by searching the Web, eventually "learning" from a user's habits and internet history to tweak its results. Soon, other tech companies began including virtual assistants into their smartphones. In 2014, Amazon announced their virtual assistant for the home, Alexa, and two years later, Google Home was released. These assistants use microphones, placed in a smart speaker, that passively listen, awaiting a voice command.

FIRST COMMERCIAL QUANTUM COMPUTER—2019

Quantum computers are a completely new kind of computer. Classical computing, used by all the computers previously described in this book, is made up of logic gates built from transistors that manipulate 1s and 0s. Quantum computing instead uses quantum bits called qubits. Qubits can hold a 1 and a 0 at the same time. The value of a qubit is manipulated using the three properties of quantum mechanics: superposition, entanglement, and interference. Superposition is the ability to hold more than one state—for example, when a coin is spinning, it is neither heads nor tails.

The first quantum computing demonstration was in 1998. It had two qubits and would work for only a few nanoseconds at a time. In 2019, IBM revealed the first quantum computer that can be utilized for science projects and commercial use outside of its research lab.

Researchers believe quantum computers will help solve problems that were previously too complex for classical computers. There is still a very long way to go for the practical use of these new supercomputers. Our ability to use quantum computing in the 2020s is very much like classical computing was in the 1940s—certainly exciting but nowhere near reaching its full potential.

THE SMART HOME—2011

Scientists at the University of Cambridge were tired of walking all the way to the main lab, called the Trojan Room, for coffee only to find an empty pot! Their solution was to set up a digital camera that faced the coffeepot. This way, they could check on the coffee status remotely. In 1993, the Trojan Room Coffee Pot went live on the internet and became the world's first webcam. This was far from a smart device, but it demonstrated the convenience of having remote access to household items.

The Nest Learning Thermostat was released in 2011 and was one of the first successful smart household products. The Nest connects to the internet and can be controlled remotely from a smartphone. The Nest also uses AI to learn a user's preferences and temperature patterns. The commercial success of the Nest kicked off an entire market of smart home appliances.

INFLUENTIAL PEOPLE

KIMBERLY BRYANT • 1967–

"JOBS IN TECHNOLOGY HAVE THE RAPIDEST RATE OF GROWTH. THE NEED FOR COMPUTER SCIENCE IS SO INCREDIBLY LARGE, AND IT'S IMPORTANT THAT GIRLS OF ALL COLORS HAVE THE OPPORTUNITY TO MOVE INTO THAT FIELD."

IN 2013, SHE BECAME A WHITE HOUSE CHAMPION OF CHANGE FOR TECH INCLUSION.

SHE'S AN ELECTRICAL ENGINEER WHO WORKED IN THE BIOTECHNOLOGY FIELD AND FOUNDED BLACK GIRLS CODE IN 2011 TO ADDRESS THE UNDER-REPRESENTATION OF BLACK WOMEN IN TECHNOLOGY.

JENSEN HUANG 1963–

"THE DISPLAY IS THE COMPUTER."

HE IS A TAIWANESE AMERICAN BUSINESSMAN AND ELECTRICAL ENGINEER.

HE CO-FOUNDED GRAPHICS-PROCESSOR COMPANY NVIDIA IN 1993. NVIDIA'S GPUS POWERED THE RISE OF 3-D GAMES AND GRAPHICS ON THE PC.

NVIDIA GREW TO BE A MAJOR DESIGNER OF SPECIALIZED MICROPROCESSORS USED IN SUPERCOMPUTERS AND GRAPHICS CARDS.

GEOFFREY HINTON 1947–

HE HAS BEEN NICKNAMED ONE OF "THE GODFATHERS OF DEEP-LEARNING AI."

HE WORKS AT THE GOOGLE BRAIN RESEARCH LAB AND AT THE UNIVERSITY OF TORONTO.

HIS WORK HELPED MAKE DEEP LEARNING MAINSTREAM AND HE HAS FURTHERED DEVELOPMENTS IN COMPUTER VISION.

"I HAVE ALWAYS BEEN CONVINCED THAT THE ONLY WAY TO GET ARTIFICIAL INTELLIGENCE TO WORK IS TO DO THE COMPUTATION IN A WAY SIMILAR TO THE HUMAN BRAIN... WE ARE MAKING PROGRESS, THOUGH WE STILL HAVE LOTS TO LEARN ABOUT HOW THE BRAIN ACTUALLY WORKS."

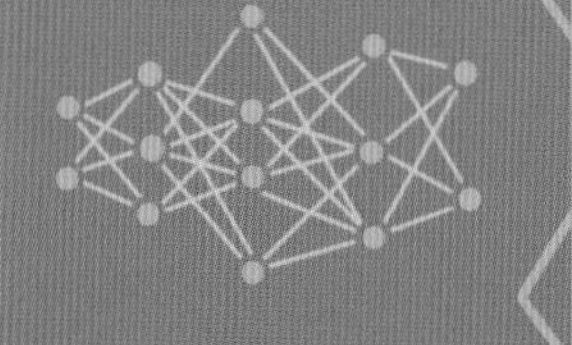

IN 1986, HINTON CO-AUTHORED AN ACADEMIC PAPER, WITH DAVID RUMELHART AND RONALD J. WILLIAMS, "LEARNING REPRESENTATIONS BY BACK-PROPAGATING ERRORS." THIS POPULARIZED THE BACKPROPAGATION ALGORITHM — A POWERFUL WAY TO TRAIN NEURAL NETWORKS.

TECH MOGULS OF THE 21ST CENTURY

HERE ARE A FEW OF THE MANY TECH BILLIONAIRES WHO HAVE CREATED AND OWN EXTREMELY PROFITABLE COMPANIES. THEY HAVE A LOT OF POLITICAL INFLUENCE AND LOBBYING POWER IN TECH, BUSINESS, PRIVACY, AND DATA COLLECTION.

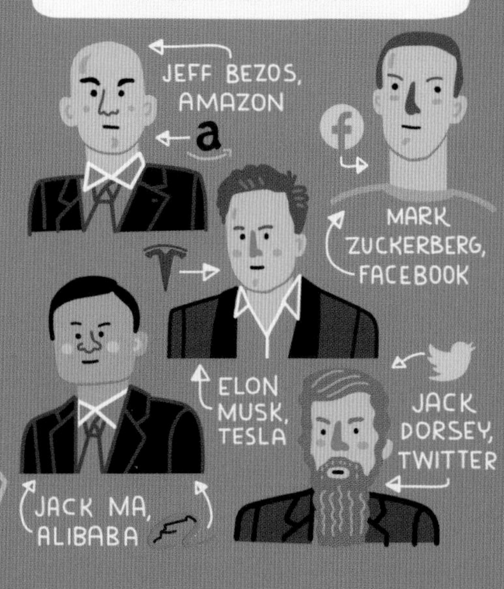

FEATURED ORGANIZATIONS

EFF

"EFF'S MISSION IS TO ENSURE THAT TECHNOLOGY SUPPORTS FREEDOM, JUSTICE, AND INNOVATION FOR ALL PEOPLE OF THE WORLD."

DIGITAL RIGHTS!

THE ELECTRONIC FRONTIER FOUNDATION 1990-

The Electronic Frontier Foundation (EFF) was started in the early 1990s to lobby the U.S. government for protection of internet civil liberties. There was confusion and a large knowledge gap around the early Web. United States law enforcement sometimes mistakenly confiscated innocent users' computers and equipment in an overzealous pursuit of early internet "hackers." The EFF worked to educate lawmakers on the new tech frontier that had changed overnight and to extend U.S. constitutional protections to the digital world.

The EFF has continued its work to protect online civil liberties and the freedom of information. Members of the EFF include computer scientists, technologists, lawyers, and activists, who work together to fight for the digital rights of internet users. Issues they focus on include a right to privacy, the ability to create and access technology, fighting against mass surveillance, and computer security.

THE INTERNET ARCHIVE 1996-

Almost all of the early internet is gone. Like silent movies from the 1920s, websites were simply trashed without a second thought when it didn't make commercial sense to preserve them.

It wasn't until 1996 when Internet Archive founders, Brewster Kahle and Bruce Gilliat, began using a "Web crawler" program to take snapshots of websites as a way to document them. The internet of the late 1990s is preserved with all of its quirks and lo-res dancing animations in the Internet Archive's Wayback Machine. This publicly available tool allows visitors to search any website URL and view what it looked like in the past.

The temporary nature of online information is still a problem. This is why the Wayback Machine continues to be an essential tool, especially for journalists. The Wayback Machine gives access to information that governments and organizations may have changed without notice or tried to hide.

Based in San Francisco, the Internet Archive is an ever-expanding digital library that continues to back up the internet many times over. The archive provides online access to millions of books, videos, audio recordings, and software.

It is the largest historical archive of computer software in the world, preserving many of the famous programs described in this book!

CHALLENGES IN A DIGITAL WORLD

Like the impact of the steam engine in the Industrial Revolution, the introduction of new computing technology has created huge shifts in how we work and organize our society.

E-WASTE

Building computers requires many nonrenewable resources such as petroleum, gold, and rare-earth elements. Many consumer electronics are designed to be unrepairable and replaced with new devices yearly. This is wasteful and ecologically irresponsible. To keep valuable materials out of landfills, electronics must be part of a future circular economy. We need consumer-friendly designs that allow tech to be repaired and upgraded so hardware can last as long as possible.

PERSONAL DATA AND PRIVACY

Thanks to smart devices, tech companies have more personal data than many users are aware of, or comfortable with. People around the globe have many different expectations of privacy that must be respected.

AUTOMATION AND LABOR

New kinds of AI and robots are being developed to ease the task of tedious mental or physical labor. In the same way that steam engines and assembly lines replaced jobs in the 1800s, there will be a reorganization of labor as automation continues to change the workforce. App companies have already disrupted traditional employment in transportation and retail, turning many positions into lower-paid gig-based work. Lawmakers are now challenged with the task of how to classify this new kind of workforce, as well as define the benefits and wages that app companies owe to employees.

TRUSTWORTHY SOURCES

On the internet you can learn about anything and everything! While this can be a great experience, there is also a lot of misinformation. The information we consume impacts our world view—whether it is actively sought or passively absorbed. It is important to always fact-check with credible sources before sharing information found online.

DIGITAL DARK AGE

Digital data may seem as if it will last forever. But with easily broken devices and a lack of physical records, it is at risk of being lost after only a few decades. A potential "digital dark age" means that archaeologists of the future might not be able to decrypt old computer files. There is no guarantee that the future machines will be able to understand the data created today. If something were to happen to digital archives, like a massive solar flare, data could be corrupted forever. Rick West, who manages data at Google, has said that "We may [one day] know less about the early twenty-first century than we do about the early twentieth century."

NET NEUTRALITY

LOADED

LOADING

STOP

STOP

Net neutrality is the principle that all websites and service on the internet must arrive to users at the same connection speed. This means that ISPs (internet service providers) are not allowed to prioritize the speed of, or access to, one website over another. Net neutrality has a great impact on the freedom of speech and freedom of enterprise. In general, Europe has the strongest net neutrality protections.

BIAS IN ALGORITHMS AND AI

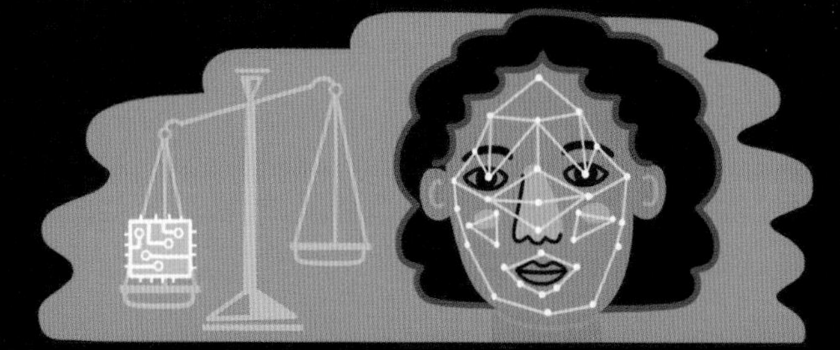

While AI is a powerful tool to abridge mental labor, it is not perfect. Algorithms will be as biased as the people who create them. This can have bad consequences when AI is used to sort through items such as job résumés and loan applications or used for facial recognition. AI is only as ethical and unbiased as the people who use it, and it can be very dangerous when used with bad intent.

DIGITAL DIVIDE

Computers and internet access have become a necessity and are especially vital in education. Many people in the United States and around the world are still unable to afford personal computers. Without proper hardware, they are unable to participate in computer-driven work or school. Many functional computers end up in landfills and e-waste centers, when they could be in the hands of those who need them. The cost of fast and reliable internet access is also a problem for millions of people. Everyone needs access to fully functional technology!

THE FUTURE

By looking at current technology and computer science research, we can guess what will come next. These are a few of the advancements that computer engineers are working toward now.

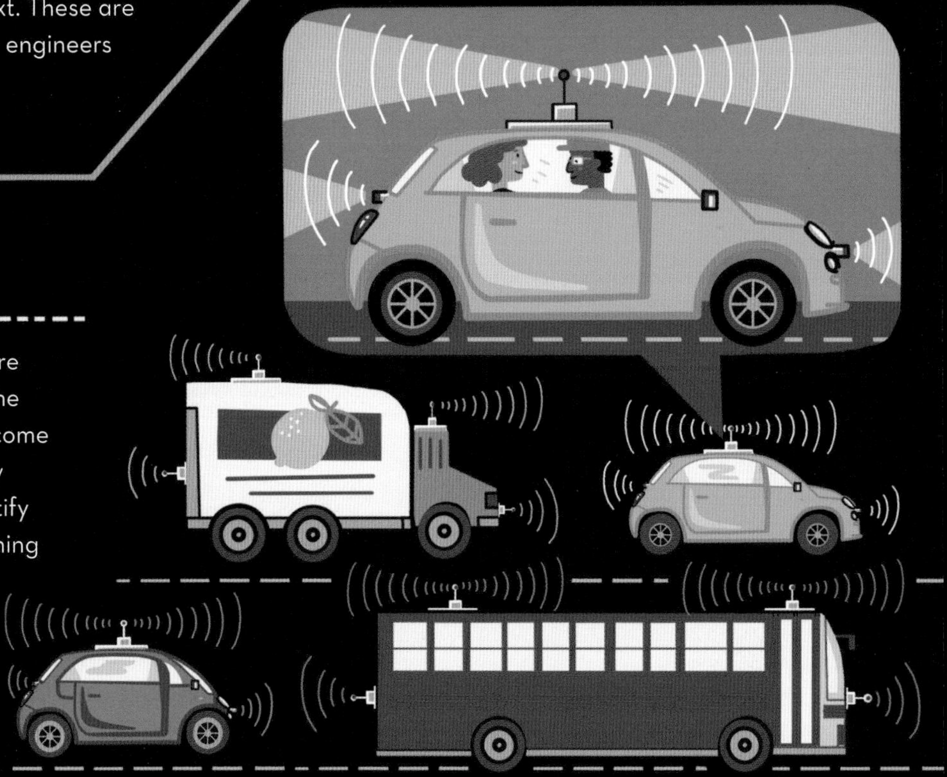

FULLY AUTONOMOUS SELF-DRIVING CARS

Today, we have self-driving cars that require human supervision. It is theorized that in the next few decades, self-driving cars will become a mainstream form of transportation. They will need very powerful computers to identify every possible hazard on the road—something that not even humans can do!

UBIQUITOUS COMPUTING

The "internet of things" is a precursor to what many computer scientists think will come next—computers everywhere and in everything! People theorize that computers will be woven into clothing, embedded in walls, and taking measurements in the air and soil. The expectation is that computers will become a mostly invisible and ubiquitous technology.

THE AI SINGULARITY

We are very far away from creating a general AI, which is an artificial intelligence that can do everything a human brain can. An AI that is even smarter than human intelligence is referred to as the "technological singularity." This is a goal for some AI researchers, but creating a super-intelligence is such a lofty objective that many experts see it more as science fiction.

BIG DATA AND HYPOTHESIS-FREE SCIENCE

Science often starts by asking questions about our universe. But what about the questions we don't even know to ask yet? Today, much of scientific work is collecting and analyzing data. Many theorize that future computers will automatically collect data using global sensors and then AI will analyze that data and find emerging patterns. So, instead of always starting with a hypothesis, scientists of the future may be presented with an observation from AI as a starting point to learn more.

Computers are arguably the greatest tool that humanity has ever created. In many ways, tools shape the limits of our imaginations. A hammer doesn't only let us strike a nail—it allows us to imagine new forms for planks of wood to take when joined together. The computer is a tool to expand our mental abilities, and it has opened new possibilities for humanity to build and dream bigger.

In the history of computers, new technology has, for the most part, been limited to the very powerful few. Computers, and even the internet, were at first only available to governments or large companies and were understood by a few specially trained people. Eventually, these technologies were "liberated" and made accessible to the general public, putting the power of computers in the hands of the people.

In the scheme of things, it has only been a brief moment in history that the power of computing has been accessible to individuals. Despite—and perhaps because of—this, it is important to be critical of how cutting-edge technology is being used. Our future doesn't have to resemble the computer world of the past, where new tech is only truly understood by the few and used by the powerful. Many of our problems can be solved when our tools—developed and used ethically and thoughtfully—work for us.

So, I ask, what will *you* do with computers and technology? What will you learn? What will you build?

"DON'T LET FEAR GET IN THE WAY AND DON'T BE AFRAID TO SAY 'I DON'T KNOW' OR 'I DON'T UNDERSTAND'— NO QUESTION IS A DUMB QUESTION."
—MARGARET HAMILTON
USA
PET
"PAYOFF WILL COME WHEN WE MAKE BETTER USE OF COMPUTERS TO BRING COMMUNITIES OF PEOPLE TOGETHER AND TO AUGMENT THE VERY HUMAN SKILLS THAT PEOPLE BRING TO BEAR ON DIFFICULT PROBLEMS."
—DOUGLAS ENGELBART
"THE FUTURE IS NOT LAID OUT ON A TRACK. IT IS SOMETHING THAT WE CAN DECIDE, AND TO THE EXTENT THAT WE DO NOT VIOLATE ANY KNOWN LAWS OF THE UNIVERSE, WE CAN PROBABLY MAKE IT WORK THE WAY THAT WE WANT TO."
—ALAN KAY

SOURCES AND RESOURCES

Interested in learning more about computer history? Here are some of the sources that I consulted while researching this book. For a full list of sources, visit my website at rachelignotofskydesign.com/the-history-of-the-computer.

WEBSITES

The ACM A.M. Turing Award: amturing.acm.org

Bletchley Park: bletchleypark.org.uk

The Doug Engelbart Institute: dougengelbart.org

The ENIAC Programmers Project: eniacprogrammers.org

The Institute of Electrical and Electronics Engineers: ieee.org

The International Spy Museum: spymuseum.org

The National Inventors Hall of Fame: invent.org

The National Museum of Computing: tnmoc.org

The National Science Foundation: nsf.gov

The Pew Research Center: pewresearch.org

The United States Census Bureau: census.gov/history

MUSEUMS

Computer History Museum
1401 N. Shoreline Boulevard
Mountain View, CA 94043
computerhistory.org

Living Computers: Museum + Labs
2245 First Avenue S.
Seattle, WA 98134
livingcomputers.org

BOOKS

Barnes-Svarney, Patricia L., and Thomas E. Svarney. *The Handy Math Answer Book*. Canton, MI: Visible Ink Press, 2012.

Boyer, Carl B., and Merzbach, Uta C. *A History of Mathematics*. 3rd ed. Hoboken, NJ: Wiley, 2010.

Campbell-Kelly, Martin, William Aspray, Nathan Ensmenger, and Jeffrey R. Yost. *A History of the Information Machine*. 3rd ed. The Sloan Technology Series. Milton Park, UK: Routledge, 2013.

Ceruzzi, Paul E. *Computing: A Concise History*. The MIT Press Essential Knowledge Series. Cambridge, MA: The MIT Press, 2012.

Evans, Harold. *They Made America: From the Steam Engine to the Search Engine: Two Centuries of Innovators*. New York: Little, Brown and Company, 2004.

Freiberger, Paul, and Michael Swaine. *Fire in the Valley: The Birth and Death of the Personal Computer*. Raleigh, NC: The Pragmatic Bookshelf, 2014.

Garfinkel, Simson L. *The Computer Book: From the Abacus to Artificial Intelligence, 250 Milestones in the History of Computer Science*. New York: Sterling, 2018.

Igarashi, Yoshihide, et al. *Computing: A Historical and Technical Perspective*. Boca Raton, FL: CRC Press, 2014.

Lam, Lay Yong, and Ang Tian Se. *Fleeting Footsteps: Tracing the Conception of Arithmetic and Algebra in Ancient China*. Singapore: World Scientific Publishing Company, 2004.

McCullough, Brian. *How the Internet Happened: From Netscape to the iPhone*. New York: Liveright, 2018.

Seife, Charles, and Matt Zimet. *Zero: The Biography of a Dangerous Idea*. London: Penguin Books, 2014.

Walsh, Toby. *Android Dreams: The Past, Present and Future of Artificial Intelligence*. London: C Hurst & Co Publishers Ltd, 2017.

Wozniak, Steve, and Gina Smith. *iWoz: Computer Geek to Cult Icon: How I Invented the Personal Computer, Co-Founded Apple, and Had Fun Doing It*. New York: W.W. Norton & Co., 2006.

ACKNOWLEDGMENTS

First, I would like to thank my husband and business partner, Thomas Mason, for his help with this project. Not only was he my research assistant, who was consulted heavily, but he was also one of my inspirations to write the book. In the early days of dating, our apartment was filled with heaps of old electronics, vintage computers, and vacuum-tube calculators that he would fix and resell to help pay for college. The best of this "junk" has become prized treasures in our home and part of our own vintage computer collection.

THOMAS PLAYING CHESS AGAINST A PDP-8 DURING OUR VISIT TO LIVING COMPUTERS MUSEUM + LABS.

This book is possible because of my team at Ten Speed Press. Thank you to Kaitlin Ketchum for being my amazing editor. She is a champion of my projects, and her support means the world to me. Behind the scenes are a ton of people to thank, including my copyeditor Dolores York, fact checker Mark Burstein, proofreaders Lisa DiDonato Brousseau and Mikayla Butchart; senior managing editor, Doug Ogan; production editor, Sohayla Farman; designer, Chloe Rawlins; and Dan Myers and the entire Ten Speed production team. A big shout-out to Windy Dorresteyn, Monica Stanton, and Natalie Yera, the marketing and publicity team.

A huge thank-you to my literary agent, Monica Odom. She continues to help my book dreams come true!

ABOUT THE AUTHOR

RACHEL IGNOTOFSKY is a *New York Times* bestselling author and illustrator, based in California. She grew up in New Jersey on a healthy diet of cartoons and pudding. She graduated from Tyler School of Art and Architecture's graphic design program in 2011. Her work is inspired by history and science. She believes illustration is a powerful tool that can make learning exciting, and she has a passion for taking dense information and making it fun and accessible.

SOME OF THE VINTAGE COMPUTERS IN MY COLLECTION.

SEE MORE OF RACHEL IGNOTOFSKY'S
BOOKS AND ART
WOMEN IN SCIENCE
WOMEN IN SPORTS
WOMEN IN ART
THE WONDROUS WORKINGS OF PLANET EARTH
WHAT'S INSIDE A FLOWER?
I LOVE SCIENCE JOURNAL
THE WONDROUS WORKINGS OF SCIENCE AND NATURE COLORING BOOK
POSTCARDS AND PUZZLES
FOR MORE VISIT
RACHELIGNOTOFSKYDESIGN.COM
@RACHELIGNOTOFSKY

INDEX

A

abacuses, 18–23, 26–27
Adobe, 107
AI (artificial intelligence), 16, 35, 43, 51, 101, 108–9, 112, 114, 115, 117
Aiken, Howard, 43, 46, 50, 62
AIM (AOL Instant Messenger), 88
Alcorn, Al, 14
Aldrin, Buzz, 57
Alexa, 110
algorithms
- bias in, 115
- social media and, 107

Alibaba, 112
al-Jazari, Ismail, 23
Allen, Paul, 68, 69, 74, 99
Alphabet, 102
AlphaGo, 16
Altair 8800, 64, 68, 69, 70
Alto, 64, 71, 81
Amazon, 94, 110, 112
Analytical Engine, 30, 33, 36, 38
Anderson, Tom, 95
Andreessen, Marc, 93
Android, 104
Antheil, George, 47
Antikythera mechanism, 21, 25
Antonelli, Kathleen, 51
AOL (America Online), 88, 92, 107
Apollo Guidance Computer (AGC), 57, 61, 63
Apple
- Apple I, 70, 75
- Apple II, 64, 66, 67, 70, 72, 75, 80
- AppleTalk, 82
- App Store, 105
- history of, 70, 75, 81, 87
- iPad, 87
- iPhone, 66, 87, 100, 104–5, 110
- iPod, 87, 88, 91, 97
- iTunes, 91
- LaserWriter, 82
- Lisa, 82
- logo for, 70
- Macintosh, 77, 81, 82, 85, 86, 87

applications, 105
APT (Automatically Programmed Tool), 17
Armstrong, Neil, 57
ARPANET (Advanced Research Projects Agency Network), 55, 56, 73, 79, 92
artificial neural networks (ANNs), 100, 109
ASIMO, 17
Asimov, Isaac, 44
astrolabes, 18, 22
Atanasoff, John, 45
Atari, 14, 75
AutoCAD, 76

B

Babbage, Charles, 28, 30, 32–33, 36, 38, 46, 118
backlinking, 94
Backus, John, 55
Baer, Ralph H., 14
banner ads, 95
Bardeen, John, 48
Bartik, Jean, 51
batch processing, 58
Bel Geddes, Norman, 45
Bell, Alexander Graham, 30
Bell Labs, 42, 48, 55, 66, 71, 83, 110
Berners-Lee, Tim, 90, 92, 93, 99
Berry, Clifford, 45
Bezos, Jeff, 112
big data, 108, 117
Bina, Eric, 93
binary code, 9, 10, 38
bit, 12, 42
Bitcoin, 103
Black Americans, contributions of, 46, 57
Blackberry, 97, 104
Bletchley Park, 43, 44, 45, 50, 51
Blue Boxes, 75
Bluetooth, 8, 47, 101
bombes, 40, 44
Boole, George, 10, 31, 38
Boolean algebra, 10, 29, 31, 38
Brathwaite, Richard, 30
Brattain, Walter, 48
Braun, Karl Ferdinand, 30
Bricklin, Dan, 70
Brin, Sergey, 94
broadband, 96, 101
Broadhurst, Sidney, 45
browsers, 90, 93
Bryant, Kimberly, 112
bulletin board system (BBS), 81
Bush, Vannevar, 42, 46, 50, 63
Bushnell, Nolan, 14
byte, 12

C

Cailliau, Robert, 90, 92
calculators, mechanical, 33
Cambridge, University of, 111
cameras, digital, 70, 88
Catmull, Edwin, 70, 83, 86
CDC (Control Data Corporation), 55
CD-ROMs, 78
CDs, 78
cellphones, 71, 91, 97
CERN (European Organization for Nuclear Research), 92, 93, 99
CGI (computer-generated imagery), 83, 94
Chandler, William, 45
Clairaut, Alexis Claude, 31
Clark, Jim, 93
Clarke, Joan, 45
cloud computing, 13, 100, 107
Coffee, William, 46
Cold War, 42, 47, 53, 56
Collins, Michael, 57
Colossus, 40, 41, 43, 44, 45, 48, 50
command line, 9
Commodore, 64, 66, 70, 77, 84, 86
Community Memory, 66, 68
Compaq, 80
compilers, 54
computer bugs, 43
computer chips, 11, 54, 60
computers
- access to, 89, 115, 118
- components of, 8
- definition of, 6, 46
- "first," 45, 46–47
- first use of word, 30
- future of, 116–17
- human, 29, 30, 32, 33, 51, 57
- mini-, 55
- personal, 65–71, 77, 80, 83
- quantum, 100, 110
- smallest, 103
- super-, 56, 90, 103, 105
- technological convergence and, 9
- as tools, 7, 89, 118
- ubiquitous, 116

Computer Space, 69
Cooper, Martin, 71
CPU (Central Processing Unit), 8
Cray, Seymour, 69

Cray supercomputers, 69, 83
cryptocurrency, 103
CTR (Computing-Tabulating-Recording), 31, 39
cyberspace, 81

D

DARPA Grand Challenge, 17
DAW (digital audio workstation), 85
DEC (Digital Equipment Corporation), 55
Deep Blue, 16
deep learning, 100, 109, 112
Dell, 80
Dendral, 16
Difference Engine, 28, 32–33, 36, 38
Differential Analyzer, 46
digital dark age, 115
digital divide, 115
Digital Research, 75
DJI Phantom, 17
DNA, as data storage, 104
Donkey Kong, 87
Dorsey, Jack, 112
dot-com bubble and crash, 90
DRAM (dynamic random-access memory), 13
drones, 17
Dynabook, 74, 106

E

Easley, Annie, 57
eBay, 94, 98
Eckert, J. Presper, 46, 49, 58
e-commerce, 94
Eisenhower, Dwight D., 6
Electric Pencil, 67
Electro, 17
Electronic Frontier Foundation (EFF), 113
ELIZA, 16
email, 73
EMCC (Eckert-Mauchly Computer Corporation), 58
Engelbart, Douglas, 54, 62, 63, 81, 119
ENIAC (Electronic Numerical Integrator and Computer), 43, 46–47, 48, 49, 51, 58
Enigma machines, 40, 44, 50
Estridge, Don, 80
Ethernet, 71
Everett, Robert, 56
e-waste, 114
expansion bus, 8

F

Facebook, 95, 107, 112
Faggin, Federico, 72, 74
Fairchild Semiconductor, 60, 62
Fedex, 109
Felsenstein, Lee, 68, 74
Fire Control, 46
flash memory, 13, 79
Fleming, John Ambrose, 31
Fletcher, Jonathan, 94
floppy disks, 13, 72
Flowers, Tommy, 44, 45, 50
Folding@home, 103
Forrester, Jay, 56
FORTRAN, 55
Frankston, Bob, 70
French, Gordon, 68
Friendster, 95
Futureworld, 70

G

Gagarin, Yuri, 57
Gates, Bill, 69, 74, 80, 93, 99
General Motors, 17
GeoCities, 88, 95
Gernell, François, 67
Gibson, William, 81
gigabyte (GB), 12
Gilliat, Bruce, 113
Goldwasser, Shafi, 98
Google, 16, 94, 102, 106, 110. *See also* Alphabet
GPU (Graphics Processing Unit), 8
GRiD Systems, 79, 84
GUI (Graphic User Interface), 9, 62, 71, 77, 78, 81, 82
Gun Fight, 67

H

Halley's Comet, 31
Hamilton, Margaret, 61, 63, 119
Hanrahan, Patrick, 86
hard disk drives, 13
hardware, 8
Harvard Mark I, 11, 40, 41, 43, 45, 46, 48, 50, 62
Harvard University, 69, 74, 95
hashtags, 103
Hawkins, Jeff, 79
Hazen, Harold Locke, 46
Heathkit, 81
Hero of Alexandria, 23
Herschel, John, 32
Hewlett, Bill, 42
High-Performance Computing Act, 90
Hinton, Geoffrey, 112
Holberton, Frances "Betty," 51
Hollerith, Herman, 34, 35, 37, 39
Holocaust, 41
Homebrew Computer Club, 68, 70, 80
Honda, 17
Hopper, Grace, 43, 54, 62
HP (Hewlett-Packard), 42, 68, 71, 75
HTML (Hyper Text Markup Language), 92
Huang, Jensen, 112
Hull, Charles "Chuck," 78

I

IBM (International Business Machines)
- competitors of, 59
- corporate culture of, 35, 39
- Deep Blue, 16
- floppy disks, 72
- history of, 31, 35, 39
- Microsoft and, 74, 80, 99
- Model 1401, 59
- PC, 76, 77, 78, 80
- quantum computer, 110
- Simon, 104
- System/360, 52, 59
- Type 77 Collator, 34
- Watson, 16

Ibuka, Masaru, 62
ImageNet, 16
Industrial Revolution, 29, 30, 32, 34–35
Instagram, 107
integrated circuits (ICs), 11, 54, 60
Intel, 62, 64, 67, 69, 72, 74, 75, 80
internet
- dial-up vs. broadband, 96
- global use of, 103
- history of, 55, 81, 92, 93, 95
- misinformation on, 114
- of things, 111, 116
- ubiquity of, 106
- Web vs., 92

Internet Archive, 113
Internet Explorer, 90, 93
iPad, 87
iPhone, 66, 87, 100, 104–5, 110
iPod, 87, 88, 91, 97
I, Robot, 44
iTunes, 91
Iwatani, Toru, 15

J

Jackson, Mary, 57
Jacquard, Joseph Marie, 36
Jacquard loom, 36
Jobs, Steve, 68, 70, 75, 78, 81, 82, 83, 87, 104
Johnson, Katherine, 57
JPEG (Joint Photographic Experts Group) standard, 91
JumpStation, 94

K

Kahle, Brewster, 113
Kakehashi, Ikutaro, 86
Kare, Susan, 86
Kasparov, Gary, 16
Kay, Alan, 74, 106, 119
Kelvin, Lord, 33
Kennedy, John F., 57
Kilby, Jack, 60
Kildall, Gary, 75
kilobyte (KB), 12
Kindle, 104
Kleinrock, Leonard, 55
Kodak, 88

L

Lalande, Joseph Jerome, 31
Lamarr, Hedy, 47
LaserDiscs, 78
LaserWriter, 82
Lawson, Jerry, 68
Lebombo bone, 20
The Legend of Zelda series, 87
Leibniz, Gottfried Wilhelm, 32
Lepaute, Nicole-Reine, 31
Li, Fei-Fei, 16
Linux, 91, 98, 104
Lisa, 82
Liskov, Barbara, 98
logic gates, 10, 11, 31
Lorenz ciphers, 44
Lovelace, Ada, 36, 38
Lucasfilm, 78, 83

M

Ma, Jack, 112
machine code, 9
Macintosh, 77, 81, 82, 85, 86, 87
magnetic-core memory, 13, 52, 56
magnetic tape, 13, 52
Manchester, University of, 48, 55
Manhattan Project, 43, 47, 50
Markkula, Mike, 70
Masuoka, Fujio, 79
Mauchly, John, 46, 49, 58
megabyte (MB), 12
Meltzer, Marlyn, 51
memex, 42
memory, 12–13
MESM (Small Electronic Computing Machine), 53
Messina, Chris, 103
Micali, Silvio, 98
Michigan, University of, 103
Michigan Micro Motes (M3s), 103
Micral N, 67
microprocessors, 67, 72, 74
Microsoft
- history of, 69, 74, 99
- IBM and, 74, 80, 99
- Internet Explorer, 90, 93
- Windows, 90, 93, 99
- Word, 83
- Xbox, 15

MIDI (Musical Instrument Digital Interface), 76, 85, 86
minicomputers, 55
MIT, 13, 14, 16, 56, 58, 60, 61, 63, 98
MITS, 69
Miyamoto, Shigeru, 87
Moore, Fred, 68
Moore, Gordon, 54, 62
Moore's Law, 54
Morita, Akio, 62
Morse code, 32
Mosaic, 90, 93
motherboards, 8
The Mother of All Demos, 54, 63
Motorola, 71
MP3 players, 97
Musk, Elon, 112
Myspace, 95

N

Napier, John, 30
NASA (National Aeronautics and Space Administration), 56, 57, 61, 63, 84
NCR (National Cash Register) Corporation, 39
Nepohualtzintzin, 26
NES (Nintendo Entertainment System), 15, 79, 87
Nest Learning Thermostats, 111
net neutrality, 115
Netscape, 93
neural machine translation (NMT), 102
neural networks, 108, 109, 112
NeXT, 82, 87
Nintendo, 15, 76, 79, 83, 87
NLS (oN-Line System), 63
Nokia Communicator, 91, 97, 104
Northwestern University, 109
Noyce, Robert, 60, 62
NSFNET (National Science Foundation Network), 79, 81, 92, 93
numerals, 20, 21, 23
Nvidia, 15, 112

O

Omidyar, Pierre, 98
ON/OFF switches, 10–11
optical storage, 13
Oregon Trail, 14
Orwell, George, 42, 82
OS (Operating System), 9
Osborne, Adam, 74
Osborne Computer Company, 74
Oughtred, William, 30

P

Packard, David, 42
packet switching, 55
Pac-Man, 15
Page, Larry, 94
PalmPilot, 79, 88, 97
pantelegraph, 34
Patterson, John Henry, 39
PDP-1, 55, 60
Pennsylvania, University of, 46
peripheral devices, 8
personal assistants, 97
Photoshop, 93
Pixar, 78, 83, 86, 87
pocket devices, 97, 106
Pokémon Go, 15
Pong, 14
pop-up ads, 91
ports, 8
power glove, 81
power supplies, 8
privacy, 108, 114
programming languages, 9
programs, 9, 36
punch cards, 13, 28, 34, 36, 37, 56, 58

Q

quantum computers, 100, 110
quipus, 24

R

Rackoff, Charles, 98
radar, 48
Radio Shack, 66, 70
RAM (random access memory), 8, 12
Raspberry Pi, 102
RDA (Rockefeller Differential Analyzer), 46
relay switches, 11
Remington, 34, 58
RenderMan, 83, 86
Ritchie, Dennis, 66
Roberts, Ed, 69
robots, 17, 35, 44, 81, 94, 114
Roland Corporation, 86
ROM (read-only memory), 12
Roomba, 17
Russell, Steve, 60

S

SAGE (Semi-Automatic Ground Environment), 54, 56
Salamis Tablet, 18, 20
Sanger, Larry, 96
Sasson, Steven, 70
science, in the future, 117
scytale, 24
search engines, 94
self-driving vehicles, 100, 109, 116
semiconductors, 30
Shakey, 17
Shannon, Claude, 10, 31, 42, 50
Shepard, Alan, 57
Shima, Masatoshi, 72
Shrayer, Michael, 67
Silicon Valley, 42
Silk Road, 21
singularity, technological, 117
Siri, 110
Sketchpad, 62
slide rule, 30
smart devices, 111
smartphones, 7, 9, 97, 98, 101, 104–6
Smith, Alvy Ray, 83
Smith, David, 86
Smith, Earnest C., 57
social media, 95, 106, 107, 109
Social Security Act of 1935, 34
software, 9, 107
Sony, 62
soroban, 27
Space Invaders, 71
Space Race, 57
Spacewar!, 14, 60
spam, 31, 73
Speak & Spell, 69
spreadsheets, 67, 70
SRI International, 17, 54
Stanford University, 17, 68, 94, 103
Star Trek, 52, 68, 97
Stibitz, George, 42, 46
Stockham, Thomas, 85
storage, 8, 12–13
suanpan, 18, 21, 27
supercomputers, 56, 69, 90, 103, 105
Super Mario Bros., 87
Sutherland, Ivan, 15, 62
Synaptics Inc., 74
Systrom, Kevin, 107

T

tablets, 74, 79, 106
Tabulating Machine Company, 35, 39
Taito, 71
tchoty, 27
Teitelbaum, Ruth, 51
telegraph, 28, 31, 38
telephones, 30. *See also* cellphones; smartphones
teletypes, 59
terabyte (TB), 12
Tesla, 112
Thompson, Kenneth, 66
3-D printing, 78
time sharing, 58
Tin Toy, 78, 83
Tomlinson, Ray, 73
Torres y Quevedo, Leonardo, 38
Torvalds, Linux, 98
Toshiba, 79
Toy Story, 83, 88
TRADIC, 55
Tramiel, Jack, 84, 86
transistors, 11, 48
Trojan Room Coffee Pots, 111
Turing, Alan, 43, 44, 45, 50, 51, 118
Turing machine, 51
Turing Test, 43, 51
Twitter, 103, 112
typewriters, 34

U

Unimate, 17
United States Census, 28, 34, 37, 39, 58
UNIVAC (UNIVersal Automatic Computer), 6, 51, 52, 54, 58, 62
UNIX, 66

V

vacuum tubes, 11, 31, 40
video games, 14–15, 87
virtual assistants, 110
virtual reality (VR), 15, 100
VisiCalc, 67, 70, 72

W

Wales, Jimmy, 96
Wang, An, 74
Wang Laboratories, 64, 74
Watson, 16
Watson, Thomas J., Jr., 39
Watson, Thomas J., Sr., 39
Wayback Machine, 113
Web
  history of, 89, 90–91, 92, 95, 99
  internet vs., 92
  2.0, 91, 95, 96, 101
webcams, 111
Weizenbaum, Joseph, 16
West, Rick, 115
Westinghouse Electric, 17
Whirlwind, 13, 14, 56
Wi-Fi, 8, 47, 101
Wikipedia, 96
Windows operating system, 90, 93, 99
Wing, Jeannette, 98
Woods, Granville, 38
word processing, 67, 83
World of Warcraft, 15
World War II, 6, 39, 41, 44, 46, 47, 48, 51, 53
World Wide Web (W3) Consortium, 99
Wozniak, Steve, 68, 70, 75
WYSIWYG (What You See Is What You Get), 71, 82

X

Xbox, 15
Xerox PARC, 63, 71, 74, 78, 81, 82, 83

Y

Yamaha, 76
YouTube, 95, 108

Z

zero, 20, 21
Zeus, Konrad, 45
Zilog, 74
Zuckerberg, Mark, 95, 112

Published in the United States by Ten Speed Press, an imprint of Random House, a division of Penguin Random House LLC, New York.
TenSpeedPress.com
RandomHouseBoooks.com

Ten Speed Press and the Ten Speed Press colophon are registered trademarks of Penguin Random House LLC.

Library of Congress Control Number: 2021950386

Hardcover ISBN: 978-1-9848-5742-2
eBook ISBN: 978-1-9848-5743-9

Printed in Italy

Editor: Kaitlin Ketchum
Production editors: Doug Ogan and Sohayla Farman
Designer: Chloe Rawlins
Typeface: House Industries' Neutraface Text
Production manager: Dan Myers
Copyeditor: Dolores York
Proofreaders: Lisa DiDonato Brousseau and Mikayla Butchart
Indexer: Ken DellaPenta
Fact checker: Mark Burstein
Publicist: Natalie Yera
Marketer: Monica Stanton

10 9 8 7 6 5 4 3 2 1

First Edition

The paper of this book is FSC® certified, which assures it was made from responsible sources.